PENGUIN BOOKS

SAVING THE PLANET

Nik Nazmi Nik Ahmad has been the Malaysian minister of natural resources and environmental sustainability since December 2023. From 2022 to 2023, he was the minister of natural resources, environment, and climate change. He is also the member of Parliament for Setiawangsa, Kuala Lumpur, and the vice president of the People's Justice Party (KEADILAN).

He began his political career as the Selangor state assemblyman for Seri Setia in 2008. Nik Nazmi served in the constituency for two terms. He was appointed as political secretary to the Selangor Chief Minister Abdul Khalid Ibrahim from 2008 to 2010.

In 2013, Nik Nazmi was appointed as the deputy speaker of the Selangor State Assembly, before being appointed as Selangor state EXCO (state minister) for education, science, technology, and human capital development from 2014 to 2018.

In 2018, Nik Nazmi won the Setiawangsa parliamentary constituency. As an MP, Nik Nazmi chaired the Parliamentary Special Select Committee on Defence and Home Affairs.

In the party, he held the position of communications director, KEADILAN youth leader, PH Coalition youth leader, KEADILAN chief organizing secretary, and KEADILAN vice president. He subsequently defended Setiawangsa in 2022 before being appointed as a cabinet minister.

Nik Nazmi co-founded the Mentari Project, an educational programme in a low-cost housing area in 2007, as well as community football club Setiawangsa Rangers FC in 2018.

He has written several books including *Malaysian Son: A Progressive's Political Journey in the Heart of Southeast Asia* and *Moving Forward: Malays for the 21st Century*.

Nik Nazmi studied at the Malay College Kuala Kangsar (MCKK) and read law at King's College London. He lives with his family in Petaling Jaya.

ALSO BY NIK NAZMI NIK AHMAD

Malaysian Son: A Progressive's Political Journey in the Heart of Southeast Asia (2022)

ADVANCE PRAISE FOR *SAVING THE PLANET*

'I am firm in my belief that this book will be an invaluable resource across the board, from the casual reader keen to explore the topic to those seeking to understand environmental issues from a Malaysian perspective to aspiring politicians on how, like the author has, to master unfamiliar but essential policy spheres like climate change. I trust that it will inspire [. . .] young and old alike to commit themselves to the cause of climate action. Let us never lose sight of the fact that we only have this one planet and that it is our responsibility to preserve it.'

—Anwar Ibrahim, prime minister of Malaysia

'Nik Nazmi's book is conveniently compact and highly readable. The book is a useful reference for both general and specialist readers. The title *Saving the Planet* is grand, so grand I was a little suspicious when first approaching the book. There are too many books with big ideas on environmental sustainability, which oversimplify the fascinating complexity of Mother Nature and disregard the implications of their prescriptions on ordinary people for many of whom life remains a daily struggle. Nik Nazmi's book is nothing of that sort.

'In every chapter—in every section—he zooms out for a higher view and zooms back in on the hard reality of human society. His love for the *tanah air* of Malaysia suffuses the pages but without any hint of a narrow nationalism, which ignores the interests of others in the region and the world. He is a political leader who is grounded in the concerns of the small man without losing sight of the big picture. Saving the planet means saving the common man. Without over-making the point, Nik Nazmi argues that our relationship with Nature, which is our mother, which therefore connects us all as brothers and sisters, is also a moral challenge about which there is much to learn from the great religions.'

—George Yeo, former foreign minister of Singapore

'"Then We made you successors in the land to observe how you would act." (Quran 10:14).

'The Islamic tradition emphasises the stewardship of the children of Adam (peace be upon him) on Earth, and of the biggest responsibilities mandated in this divinely established stewardship is that of the preservation of the environment. *Saving the Planet* by Nik Nazmi is a profound and timely work that expands on Malaysia's unique geopolitical positioning, one that allows it to be a model of sustainability in the region and a bulwark against climate change.'

—Dr Omar Suleiman, founding president of Yaqeen Institute of Islamic Research

'*Saving the Planet* is a thoughtfully written book that enriches the understanding of the history of environmental related policies in Malaysia. I particularly enjoyed the author's personal reflections, interspersed with delightful snippets of local culture, taking us on a vivid journey through the eyes of a young politician. I was struck by the honesty in acknowledging errant policies yet hope for the future. Nik Nazmi poignantly points out that in the end it is about political will and the willingness of its citizens to adapt to the current planetary crisis and ultimately, this will impact our health and well-being. An invaluable read for all concerned and interested in understanding the planetary crisis and how we must now act.'

—Dr Jemilah Mahmud, executive director of Sunway Centre for Planetary Health

'*Saving the Planet* is an important and unique critique of how Malaysia is and should be dealing with climate change. In this book, Nik Nazmi presents a holistic and comprehensive approach based on Malaysia's economic and trading economic foundations while also insisting on the need to engage with religious leaders to promote a more environmentally friendly and progressive narrative amongst Malaysians noting their religious and cultural identities.

'An important book, with some very important insights.'

—Jamal Elshayyal, Al Jazeera Media Network

'As Malaysia strides towards environmental resilience, *Saving the Planet* offers a compelling vision for a sustainable Malaysia, it emerges as a beacon of progressive thought and action that resonates deeply with indigenous communities as stewards of some of the world's oldest rainforests. Through the lens of Malaysia's unique biodiversity and environmental challenges, he [Nik Nazmi] underscores the urgent need for sustainable development and climate justice. This book is not just a call to action; it is a roadmap designed for policymakers and citizens alike, showcasing Malaysia's leadership in tackling global environmental crises. A crucial read for anyone committed to the future of our planet.'

—Adrian Banie Lasimbang, former senator, indigenous rights activist, and renewable energy advocate

'Written with authenticity, and sprinkles of humour. A must-read for all Malaysians, especially for those passionate to drive sustainability. This book is also an excellent introduction to the multidisciplinary field of environmental politics and the force of climate change in intellectual and political life. What I really love is how Nik delivers views in his position as a politician and NRES minister about how Malaysia can head toward a Net Zero future—across energy, biodiversity and other areas. With an expert grasp of the material, the science and technology are also clearly explained to bring the reader along. All-in-all it is a brilliant book for anyone trying to get a good beginning into the politics of the environment.'

—Dr Yasmin Rashid, former chairman of Malaysian Environmental NGOs coalition

'It's a powerful thing when the discourse of the climate crisis is understood from the Global South perspective. And this book is just that! It includes a history of events that set the stage especially for the Malaysian reader [for] the local to global context of the country's relations to the environment and economic spheres. It urges us to reconcile these spheres for the present and the future, while sharing much of the commendable work taken today by various sectors committed to climate solutions.

'This book in the end is an attempt to appeal to the global community . . . that we need to work together for our people and planet's survival. As a local civil society campaigner, this is a welcoming appeal from a ministerial platform because it is only when all right-holders are given the space indiscriminately to add to this global discourse, can we begin to see a just and equitable transition towards a sustainable future for us all.'

—Celine Lim, managing director of
Sarawak-based NGO, SAVE Rivers

'Our burning planet is the existential challenge of our times, one that needs stitched-up and multidisciplinary answers. This is a rare discourse by a writer who covers many required bases: a thinker, an activist and now a minister of environment. A must read for all who want to be a part of the solution.'

—Azman Mokhtar, chairman of
INCEIF University

'A frank and personal account of the environmental issues that concern us all. Nik Nazmi confirms with this book that he is a public figure whose voice must be heard. His grasp of the challenges, many of them shared by Brazil and Malaysia as tropical and megadiverse countries, is strong. Nik Nazmi addresses the damage caused by

humans but also his faith in our capacity to save the Earth. As he reminds us: "We do not have anywhere else to go."'
—Ary Quintella, author and ambassador of Brazil to Malaysia

'Nik lays the landscape for Malaysians and geo-enthusiasts, just as he determinedly pursues understanding himself. Detailed and encompassing, with an openness to all the different lenses that a politician and ally who's come into great responsibility requires—this read brings you on an iterative journey of defining Malaysia's place in the climate crisis.'
—Melissa Tan, climate action and sustainability advocate

'Drawing on years of experience in the corridors of Malaysian politics and environmental advocacy, *Saving the Planet* examines the many dimensions impacting current climate challenges in Malaysia. Underscored by Nik Nazmi's personal anecdotes and professional insights, the book highlights the crucial balance between economic growth and environmental responsibility.

'This book serves as a compass, guiding readers through the complexities of local and international environmental landscapes while discussing pragmatic pathways toward sustainable solutions. The focus on collective action reminds us of our responsibility to safeguard our planet for future generations who will reflect on present-day decisions.'
—Farhana Shukor, Malaysia youth delegate at COP 26 to COP 28

'This book is an excellent overview of how contemporary Malaysia [. . .] envisions the complex web of environmental and ecological challenges it is currently facing and how it proposes to address them. Nik Nazmi [. . .] is well positioned, politically and intellectually, to provide a good overview of the country's biophysical environment and accordingly to propose intelligent future plans for a better environmental and ecological health for the country. The book testifies to the well-informed perspectives of its author.

'The great merit of the book lies in its inclusive and holistic approach to the solution of the environmental problem: inclusive in the sense that it seeks to rally the whole country to the environmental cause by demonstrating that the issue concerns the common interests of all citizens; and it is holistic in the sense that it takes into account all the main contributing factors to the solution of the problem, including political will, religious values, ecofriendly science and technology, and the business sector. This book is valuable to scholars, policymakers, students, and members of the public, especially those directly concerned with environmental issues.'

—Dr Osman Bakar, emeritus professor,
Al-Ghazali Chair, Institute of
Islamic Thought and Civilization

'I've known Nik Nazmi for almost twenty years, and I recall a teh tarik session in 2007 when we dreamed of a new Malaysia, able to lead on the international stage on various fronts. Nik Nazmi invites us to take a front row seat into his work as NRES minister, as well as his vision for a liveable and sustainable planet for the generations to come.

'I appreciate that Nik Nazmi, having been raised in a multicultural and multi-religious Malaysia begins his book by highlighting the centrality of our faith to taking care of God's creation. He aptly asserts that the problem at hand is a problem of man's greed and moral decay and therefore calls on religious leaders to enter the ring.

'Compact, informative and highly readable for the layman!'
—Rev Yee Siew Meng, ordained minister
of the Presbyterian Church of Malaysia

'The book takes a comprehensive look at Malaysia's position on climate and the environment. It is clear Nik Nazmi understands Malaysia and the world's historical milestones regarding the environment, using this to prepare for an uncertain future. [...] The book encapsulates his priorities while maintaining good relationships with different stakeholders to reach a common cause.

'It is hard to get a politician who wants to focus on climate justice, and I think Nik Nazmi's recent work reflects the need for Malaysia to pursue it. Coming into a ministry with little experience of environmental advocacy, he stepped [...] up. Nik Nazmi listened and elevated marginalized voices, such as the youth, considered overlooked environmental issues in Malaysia, such as biodiversity loss as well as climate adaptation, and championed what most of the Global South demands out of the international community for climate action.'
—Aidil Iman, youth climate
change consultant, UNICEF

PRAISE FOR NIK NAZMI NIK AHMAD

'I am delighted to hear you share my vision reflected in the book (*Breathe*). Air quality is an issue very close to my heart, and protecting residents from the devastating consequences of air pollution is top of my agenda . . . Thank you for highlighting some of the programmes you are leading in Malaysia to address the climate emergency.'

—Sadiq Khan, mayor of London and author of *Breathe: Tackling the Climate Emergency*

'Nik Nazmi is the top performing minister in this Madani government. He is sensible, understands his brief, does not waste his time attacking the opposition without any justification and gets his work done. Heads and shoulders above the rest, only problem is he supports Liverpool. Oh well, can't have everything.'

—Khairy Jamaluddin, former Malaysian minister of science, technology and innovation and minister of health

'Nik Nazmi Nik Ahmad announced the lifting of the blanket subsidy on electricity on the commercial sector barely two weeks after he was sworn in as a cabinet member. The change in outlook for energy transition is palpable [. . .] Nik Nazmi also lifted the renewable energy export ban, shook up the Energy Commission's board, reviewed the third-party access framework for the power sector [. . .] He was also developing a new water tariff mechanism.'

—*The Edge*

Saving the Planet
Climate and Environmental Lessons from Malaysia and Beyond

Nik Nazmi Nik Ahmad

PENGUIN BOOKS
An imprint of Penguin Random House

PENGUIN BOOKS

Penguin Books is an imprint of the Penguin Random House group of companies whose addresses can be found at global.penguinrandomhouse.com

Published by Penguin Random House SEA Pte Ltd
40 Penjuru Lane, #03-12, Block 2
Singapore 609216

First published in Penguin Books by Penguin Random House SEA 2024

Copyright © Nik Nazmi Nik Ahmad 2024

All rights reserved

10 9 8 7 6 5 4 3 2 1

The views and opinions expressed in this book are the author's own and the facts are as reported by him which have been verified to the extent possible, and the publishers are not in any way liable for the same.

Please note that no part of this book may be used or reproduced in any manner for the purpose of training artificial intelligence technologies or systems.

ISBN 9789815204865

This book is sold subject to the condition that it shall not, by way of trade or otherwise, be lent, resold, hired out, or otherwise circulated without the publisher's prior consent in any form of binding or cover other than that in which it is published and without a similar condition including this condition being imposed on the subsequent purchaser.

www.penguin.sg

This book is dedicated to my son Ilhan and his generation.

Now that I have the responsibility of not merely being his father but also someone who has a duty to the entire nation to ensure that we leave behind a liveable planet for the young, I hope he understands the time I spend at work is for a bigger cause.

Contents

List of Abbreviations	xvii
Foreword	xxiii
Preface	xxvii
Introduction	1
On the Shoulders of Giants	19
Energy Transition	29
Transforming the Water Sector	61
Liveable Cities	91
A Sustainable Environment	119
Forests and Wildlife Conservation	159
Climate Justice for All	195
Conclusion	227
Index	249
Select Bibliography	267
Acknowledgements	271

List of Abbreviations

AATHP	ASEAN Agreement on Transboundary Haze Pollution
ACE	ASEAN Centre for Energy
ASEAN	Association of Southeast Asian Nations
ASMC	ASEAN Specialised Meteorological Centre
BA	Barisan Alternatif (Alternative Front)
BCX	Bursa Carbon Exchange
BN	Barisan Nasional (National Front)
BRT	Bus Rapid Transit
CAP	Consumer Association of Penang
CATS	Conservation Assured Tiger Standards
CBAM	Carbon Border Adjustment Mechanism
CCC	Climate Change Committee
CCS	Carbon Capture and Storage
CFC	Chlorofluorocarbon
COP	Conference of the Parties (the supreme governing body of an international convention, normally referring to UNCCC)
CORSIA	Carbon Offsetting and Reduction Scheme for International Aviation
CSR	Corporate Social Responsibility
DAP	Democratic Action Party
DID	Department of Irrigation and Drainage
DOE	Department of Environment

EC	European Commission
ECO	Environmental Children's Organisation
EECA	Energy Efficiency and Conservation Act
EFT	Ecological Fiscal Transfer for Biodiversity Conservation
EIA	Environmental Impact Assessment
EPA	Environmental Protection Agency, US
EPSM	Environmental Protection Society of Malaysia
EPU	Economic Planning Unit, Prime Minister's Department
ESG	Environmental, Social, and Governance
ETS	Electric Train System
EU	European Union
FDPM	Forestry Department of Peninsula Malaysia
FDRS	Fire Danger Rating System
FELDA	Federal Land Development Authority
FPIC	Free, Prior and Informed Consent
FRIM	Forest Research Institute of Malaysia
G20	Group of 20
GBI	Green Building Index
GET	Green Electricity Tariff
GPS	Global Positioning System
HETR	Hydrogen Economy and Technology Roadmap
ICAO	International Civil Aviation Organization
ICPT	Imbalance Cost Pass-Through
IMF	International Monetary Fund
IPCC	Intergovernmental Panel on Climate Change
IPP	Independent Power Producer
IRA	Inflation Reduction Act
IRENA	International Renewable Energy Agency
IUCN	International Union for the Conversation of Nature

IWK	Indah Water Konsortium
JETP	Just Energy Transition Partnership
JICA	Japan International Cooperation Agency
KEADILAN	Parti Keadilan Rakyat (People's Justice Party)
KUTS	Kuching Urban Transportation System
LCCF	Low Carbon Cities Framework
LED	Light-emitting diode
LLN	Lembaga Letrik Negara (National Electricity Board)
LNG	Liquified Natural Gas
LRT	Light Rail Transit
METMalaysia	Malaysian Meteorological Department
MNS	Malaysian Nature Society
MOSTI	Ministry of Science, Technology and Innovation
MRT	Mass Rapid Transit
NADMA	National Disaster Management Agency
NGO	Non-Governmental Organization
NEP	New Economic Policy
NETR	National Energy Transition Roadmap
NRECC	Ministry of Natural Resources, Environment and Climate Change
NRES	Ministry of Natural Resources and Environmental Sustainability
NRW	Non-Revenue Water
OPEC	Organization of Petroleum Exporting Countries
PACOS	Partners of Community Organizations of Sabah
PADU	Central Database Hub
PAS	Parti Islam Se-Malaysia (Pan Malaysian Islamic Party)
PEFC	Programme for the Endorsement of Forest Certification
Petronas	Petroliam Nasional Berhad (National Petroleum Limited)

PH	Pakatan Harapan (Coalition of Hope)
PNBCAP	Penang Nature-Based Climate Adaptation Programme
PPBM	Parti Pribumi Bersatu Malaysia (Malaysian United Indigenous Party)
PR	Pakatan Rakyat (People's Coalition)
PRABN	National Flood Forecasting and Warning Centre
PSI	Pollutant Standards Index
PWD	Public Works Department
RE100	Global initiative among corporates committed to 100 per cent renewable electricity
RON	Research Octane Number
SEDA	Sustainable Energy Development Authority
SESB	Sabah Electricity Sdn Bhd
SMART Tunnel	Stormwater Management and Road Tunnel
SME	Small- and Medium-sized Enterprise
SPAN	Suruhanjaya Perkhidmatan Air Negara (National Water Services Commission)
ST	Suruhanjaya Tenaga (Energy Commission)
SUV	Sport Utility Vehicle
TNB	Tenaga Nasional Berhad (National electric utility corporation)
ULEZ	Ultra-Low Emission Zone
UN	United Nations
UNCCC	UN Climate Change Conference
UNCLOS	UN Conference on the Law of the Sea
UNDP	UN Development Programme
UNESCO	UN Educational, Scientific and Cultural Organization

UNFCCC	UN Framework Convention on Climate Change
UN-Habitat	UN Human Settlement Programme
UV	Ultraviolet
WCPA	World Commission on Protected Areas
WHO	World Health Organization
WMO	World Meteorological Organization
WTE	Waste-to-Energy
WWF	World Wide Fund for Nature

Foreword

In 2004, while still in solitary confinement within the dismal walls of incarceration, I received a letter from a young man whose words had the turn of phrase of someone much older. My sacking in 1998 had unleashed a political upheaval that then set in motion the Reformasi movement, which has since altered the course of our history. It galvanized Malaysians across colour, creed, and ethnicity, young and old, to take to the streets to demand justice and reform.

That young man, an avid reader and writer, with a passion for reform, is Nik Nazmi, now a cabinet minister, having planted his roots in KEADILAN as my private secretary in 2006, rising through the ranks to his current position.

As the minister of natural resources and environmental sustainability, Nik Nazmi's current offering—*Saving the Planet*—details his time in office, more reflective of intent on an array of ideals, beliefs, and agenda than a check box of achievements—and understandably so.

Undoubtedly, sustainability is at the centre of the Unity Government's agenda for Malaysia. It is at the core of everything we do, embedded in the overarching Malaysia Madani philosophy and the Madani Economy framework, as well as sectoral initiatives like the National Energy Transition Roadmap and the New Industrial Master Plan 2030.

It's a whole-of-nation approach, a refutation of the notion that climate action is a top-down initiative imposed by the state or the elite upon the rakyat and private sector. Rather, we seek to empower and involve all layers of society to ensure that the necessary measures in mitigating and adapting to climate change will not only benefit the environment but also enhance the living standards of our people via high-quality jobs in the green economy.

The key to energizing Malaysia and setting it on the right trajectory for both the climate and the population is to ensure that the inevitable transformations of our lives and livelihoods follow a path that is just, inclusive, and equitable. To my mind, this is the crux of Nik Nazmi's book.

Domestically, this entails economy-wide initiatives like targeted electricity subsidies, practical measures such as facilitating Malaysian households to adopt solar energy and be more energy efficient, ensuring SMEs—the backbone of the nation's economy—have the capacity to meet ESG standards, and building community resilience towards adverse climate impacts.

Meanwhile, the Ecological Fiscal Transfers programme recognizes the vital role of federal and state collaboration for the Malaysian environmental agenda by incentivizing state administrations to conserve our rich forest heritage and, consequently, preserve our invaluable biodiversity.

Externally, Malaysia is a firm proponent of the 'common but differentiated responsibilities' principle and consistently strives to ensure that developed countries support developing ones in implementing critical but costly climate solutions, including energy transition. While addressing climate change and environmental degradation is the responsibility of all nations, its success or failure ultimately hinges on the Global North and South cooperating rather than competing with each other. Whether this involves technology transfers, competency building, financing, or any of the other myriad hurdles, these issues must be tackled fairly and with the utmost urgency.

In tandem with material concerns and existential imperatives, Malaysia sees the sustainability agenda as central to our faiths and a reflection of our long-held values. As Nik Nazmi writes in his text, most cultures and religions have venerable traditions of living in harmony with and preserving the environment while conscientiously utilizing its resources. Malaysia is no different in this regard. We must collectively rediscover our bond with the natural world, recognizing our moral obligation to reverse the human-induced environmental degradation of our planet.

Solving the climate challenge involves not only tangible and impactful action on the part of governments, international bodies, companies both big and small, communities, and individuals—but also a foundational change of mind and heart.

As I stressed in my September 2023 speech at the UN General Assembly: 'Scientists have confirmed that the world has just experienced its hottest summer in history. We have heard Secretary-General Guterres gravely declare that "Climate breakdown has begun." Even Malaysia is seeing an increase in adverse impacts of climate change, increasing temperature, rising sea levels, intensified monsoons, and erratic weather patterns disrupting livelihoods and degrading local ecosystems. As such, we have not a moment to lose.'

—Anwar Ibrahim, prime minister of Malaysia

Preface

I started in politics early, being the youngest elected candidate, at twenty-six, during the 2008 general election. But I started writing *earlier*. My opinion pieces were first published by then-fledgling Internet news portal *Malaysiakini*, seven years prior to the 2008 election. Writing helps me organize my thoughts and holds me accountable, as people can see what my thoughts and views are. I decided to write a book challenging the identity politics of race, *Moving Forward*, in 2009 and have written several books, in English and Malay, since then. Obviously, my thoughts have evolved, partly because of the natural process of maturing views as one age but also because of changing facts and circumstances that need a different outlook. But the basic yearning to not accept things as they are and making a case for how they should be remains at my core as a progressive, reformatory policymaker and politician.

Soon after being appointed as NRECC minister in December 2022, I felt that it was important for me to write something on the environment and climate change to share what I have learned in the form of my thoughts and experiences at both NRECC and later NRES. I have tried my best to make this book accessible for the general reader.

Malaysia is, after all, a fast-developing, middle-income nation, with a successful oil and gas sector that has not only contributed immensely to our growth but also government revenue specifically. Yet, there is a strong economic case to be made for sustainability. The economy is very much based on trade and will be impacted by more stringent environmental and decarbonization requirements by governments as well as corporations globally. Malaysia also has one of the oldest rainforests in the world and is a megadiverse country. I believe it is important to show what we are doing to protect our environment and make the country more climate resilient, hopefully adding to the chorus demanding that rich nations do more to assist us.

I have found much of my work in NRECC and NRES technical, but I enjoy the intellectual challenge. I am not a scientist or an engineer but a law graduate. I was in frontline politics as a legislator for almost fifteen years. To navigate the science and economics of the electricity and water sector (during my time at NRECC) and environment and climate change, I am fortunate to have a young but solid team at the minister's office; experienced, capable and dedicated civil servants in the ministry; as well as activists, young and old, who are passionate about these causes.

I began writing this book after a few months in office as the NRECC minister, but following the one-year reshuffle, I was designated as the NRES minister, holding on, essentially, to the natural resources and environmental half of the portfolio. Thus, I draw on my time handling both portfolios. My focus during the two stints, ultimately, has been the triple planetary crisis. At the end of the day, cliché as it may sound, there needs to not only be a whole-of-government but also whole-of-society approach to deal with the existential planetary crises. In my short time in office, I have learned a lot from Malaysia and other countries about what can be done next. I hope I can offer some useful viewpoints as someone holding office from a developing, megadiverse country.

Introduction

'Do you want to be a leader that looks back in time and say that you were on the wrong side of the argument when the world was crying out for a solution?'

—Jacinda Ardern, former New Zealand prime minister

Image 1: Managed to take a selfie with Jacinda Ardern at the Earthshot Prize 2023 in Singapore

After the 2022 Malaysian General Election, I was suddenly entrusted with the heavy responsibility of being NRECC minister. Following a year in office, Prime Minister Anwar Ibrahim did a minor cabinet reshuffle, and my ministry was split into two: I held onto the NRES ministry while the energy transition and public utilities (electricity, water, and sewage) were hived off to a new, separate ministry.

On 2 December 2022, a supporter and friend, Tan Ching Meng, treated my family and me to dinner following my election victory.

Following the election, Anwar was appointed as Malaysia's tenth prime minister. His journey to the premiership was long and winding—he had been groomed as an incoming prime minister in 1993, when I was only eleven. Anwar was sacked as deputy prime minister in 1998, jailed on spurious charges, released in 2004 to lead the single-seat KEADILAN party, brought the opposition together to make significant gains four years later, imprisoned again in 2015, and then released when the PH coalition won Federal power in 2018. In 2022, after twenty-four years of battling the odds, he had finally got the top job, the first prime minister from a multiracial party in Malaysia's history.

The dinner conversation alternated between stories about our children, as is common among parents, and politics.

At around 8.00 pm, I received a call from Anwar's assistant. He passed the phone to the Prime Minister: 'I want you to handle the NRECC portfolio,' he said. 'The swearing-in will be tomorrow afternoon at Istana Negara.'

The Prime Minister was brief and sounded slightly tired. Post-election, he had to form the Cabinet for the coalition government of parties, which, just weeks ago, had been bitter, decades-long rivals. He was calling the new cabinet ministers one-by-one. I remember watching an episode of *Yes Minister* in which senior members of parliament stood by their phones, hoping for a call from the prime minister. The mobile phone has, well, allowed us to be mobile. Yet, the rest of the process felt similar, if a bit more serious.

I managed to say thank you but otherwise, I was at a loss for words. The Prime Minister had announced that a press conference to announce his Cabinet would be held around 8.15 p.m. that night, and we were all preparing to watch it on YouTube on our phones. After Anwar ended the call, as I repeated the news to my family and Ching Meng's, I found myself trying to comprehend what exactly I had been roped into. I made a mental note of NRECC being new ministry and a mouthful at that. Everyone was obviously exultant and congratulated me. It was only when my name was announced by Anwar during his press conference that it finally seemed real. I had felt excitement and despair before. I remembered the confidence going into the 2013 General Election, when the Pakatan Rakyat (People's Coalition or PR) coalition seemed certain of winning Federal power (we did not get a majority of the seats, but we did receive a majority of the popular vote). Then, there was the 2018 experience when many were certain that I would be a minister if not at least a deputy minister, as I was the PH and KEADILAN youth leader. Drawing upon these diverse experiences, I tried to manage my expectations to avoid being disappointed. *Que sera sera.*

I got back home and Googled the work of the two former ministries that had now been merged to form NRECC: the Ministry of Energy and Natural Resources as well as the Ministry of Environment and Water. I remembered thinking that now, the buck stops with me. I had to take a good, hard look at our environmental and climate policies.

The Triple Planetary Crisis and Malaysia

After I was appointed as NRECC minister, London's Mayor Sadiq Khan's book, *Breathe*, was published. As a member of the Cabinet, I particularly appreciated one quote from it: 'Tough action on climate, many politicians think, is the triple whammy: an unsolvable problem, which doesn't matter in the here and now, and which will lose you votes.'[1]

But what the countries of the world are really facing is a triple planetary crisis: climate change, pollution, and loss of biodiversity. By now, the fact that these things are actually happening to us and are backed by science should be obvious to any discerning person. Climate change has now truly reached critical levels: human induced global temperature rise has reached 1.1 degrees Celsius. During the 2015 Paris Agreement signed by 194 countries as well as the European Union (EU), signatories pledged to keep the rise in global temperature to below two degrees Celsius, preferably limiting it to 1.5 degrees Celsius. However, in 2023, it was predicted that that the world would (briefly) breach the 1.5 degrees limit within the next five years.[2]

Air pollution results in seven million people dying prematurely across the world every year.[3] Almost the entire world's population is breathing polluted air, leading to diseases such as asthma, stroke, lung cancer and heart conditions.[4] The vertebrate populations have plunged by an average of almost 70 per cent from 1970 to 2018. But if you zoom into Latin America and the Caribbean specifically, the number is a sobering 94 per cent on average.[5] All

---

[1] Khan, Sadiq, *Breathe: Tackling the Climate Emergency*. London: Hutchinson: Heinemann, 2023, p. 19.

[2] WMO, 'Global Annual to Decadal Climate Update (Target Years: 2023-2027)', 2023, https://library.wmo.int/idurl/4/66224 [Accessed July 4, 2023].

[3] WHO, '7 Million Premature Deaths Annually Linked to Air Pollution', March 25, 2014, https://www.who.int/news/item/25-03-2014-7-million-premature-deaths-annually-linked-to-air-pollution [Accessed April 11, 2024].

[4] WHO, 'Billions of People Still Breathe Unhealthy Air: New WHO Data', April 4, 2022, https://www.who.int/news/item/04-04-2022-billions-of-people-still-breathe-unhealthy-air-new-who-data [Accessed April 11, 2024].

[5] WWF, 'Living Planet Report 2022: Building a Nature Positive Society', 2022, pp. 32–33, https://wwflpr.awsassets.panda.org/downloads/lpr_2022_full_report.pdf [Accessed April 11, 2024].

of this should be a warning to my country, Malaysia, as it sits on the biologically crucial Sundaland, which is also subject to massive threats and challenges.

Malaysians are often reminded how lucky we are compared to our neighbours. Unlike Indonesia or the Philippines, we are outside the ring of fire, which means Malaysia is not located around active volcanoes and is generally out of earthquake territory. Peninsular Malaysia is shielded from the Indian Ocean by Sumatera, while Sabah is also protected from the Pacific Ocean and the strong typhoons that regularly hit the Philippines. Our unique circumstances ironically exacerbate the challenge of making the political case for climate-resilient policies. This is because the impact of climate change can be deemed to be abstract and distant, particularly to some quarters of the media that are obsessed with the 24-hour news cycle and certain politicians who only understand four- or five-year election cycles.

But Malaysia, today, is experiencing the impact of the triple planetary crisis firsthand. Most of our population lives near our over 4,600-kilometre-long coastline. We face heatwaves, peat fires, and water shortages during the dry season, resulting in the horrible annual haze. The monsoons worsen the condition with floods in low-lying areas while hillslopes suffer from terrible landslides. Countless lives are lost.[6] Hundred-year floods are happening

---

[6] Malaysia has been experiencing massive deforestation, although this has slowed down considerably. Monsoons are becoming more intense as days of heavy rainfall increase. Droughts and floods are predicted to reduce rice yields by 60 per cent. Average surface temperatures have been increasing up to a quarter degree Celsius every decade since 1970. The UN reports that the Peninsula is facing up to a half metre increase in the sea level, while the sea level rise in Sabah will exceed 1 metre by 2100. There are studies that suggest that as early as 2040, all our mangrove forests will be submerged by the sea. See Lum, Milton, 'The Effects of Climate Change in Malaysia', *The Star*, July 5, 2022, https://www.thestar.com.my/lifestyle/health/the-doctor-says/2022/07/05/the-effects-of-climate-change-in-malaysia [Accessed November 8, 2023].

more often.[7] Our coasts are plagued by erosion as the ocean swallows former villages and beaches. We have the best carbon sinks—our forests and oceans absorb more carbon than they release—that serve not only Malaysia but the planet. Yet, they, too, are under pressure from climate change, biodiversity loss, and unsustainable development. While Malaysia's overall contribution to carbon emissions is low, our per capita global emission is still above average.

In 2020, the world came to a halt in the biggest disruption since the Second World War: the Covid-19 pandemic. The extent of globalization and technological innovation brought the world closer together. Trade depended on just-in-time supply chains. Companies kept minimum stock, eliminated waste and, with the advent of the Internet and faster delivery times, could supply goods across the supply chain quickly to the final consumer. During the Covid-19 pandemic, office-oriented work culture collapsed as work-from-home became not only viable but also necessary. Malaysian schools that had been classroom-oriented had to transition to home-based learning, pressuring parents to adapt. This exposed the inequality in society by squeezing poorer families, who were already facing job losses and loss of income, with the burden of getting decent gadgets and reliable Internet connections or having to rely on slower, manual methods of learning for their children. With cheap flights and better highways and railways, tourism had become an important part of Malaysia's economy. However, with planes grounded at airports, the thriving hospitality sector came to a screeching halt.

As economies went into lockdown, carbon emissions dropped globally. Wildlife began to roam comfortably in the empty streets

---

[7] A normal measure of floods is the average recurrence interval, which is a probability. Often, a 100-year flood is described as a once-in-a hundred years flood, but it actually means there is a 1 per cent chance in any given year for the flood to happen.

of some cities while populations of flora and fauna in nature reserves thrived in the absence of tourists. A mix of factors—reduced economic activity, the La Niña phenomenon, and better management of peat land—meant that the public did not experience the usual haze in Southeast Asia. In Port Dickson—a seaside resort town that has been suffering for so long from pollution and large number of visitors due to its proximity with Kuala Lumpur—turtles and dolphins were sighted. In Perak, an elephant inspected one classroom after another in a school. In another school in Pahang, a tapir fell into a drain while students were taking exams nearby. True, this moment was brief, and as soon economies opened, we got back to our old lifestyles. But, it was arguably a glimpse at how another world was possible.

I spoke about many issues in Parliament, but admittedly, my interventions on environment and climate change were rare until I became a minister.

Before holding elected office, I was establishing an education programme for the urban poor. Later, as a member of Parliament, I started a community football club that reached out to children in low-cost housing to encourage them to study while allowing girls to play football at a higher level. So, environment and climate change were not issues people associated me with until 2022.

Yet, in a way, 'the environment' was something no urban State Assemblyperson or member of Parliament could completely avoid. I found myself called upon to advocate for or deal with everything from uncollected rubbish to residents' anxiety about the green lungs of their neighbourhoods being destroyed for the sake of 'development'.

In fact, I was involved in two major 'green' controversies during my legislative career. In Selangor, I defied my own state government to help my constituents oppose the redevelopment of a popular sports field in Kelana Jaya. In Setiawangsa, a major controversy emerged over the development of a hill, which was a green lung there, called Bukit Dinding. People were particularly anxious about the hill because it was the site of a landslide that had affected the homes of two former ministers (including one owned by my predecessor in Setiawangsa) in 2012.

I knew that my constituents, whether in Selangor or Kuala Lumpur, were increasingly concerned about 'green issues', and they were not something that just NGOs were talking about. In fact, with the right spokespersons and narrative, people can understand that the much-talked-about 'bread-and-butter issues' are not isolated from environment and climate change. I knew climate change was something we all had to worry about, contrary to the narratives being peddled by few, powerful vested interests. I also knew that sooner or later, we would have to move away from fossil fuels. This would be an interesting conundrum (to say the least) for Malaysia as a developing country as well as an oil and gas driven economy.

One principle that I have held on to since I became involved in politics as a nineteen-year-old student has been that identity politics—based on race and religion—is attractive to our baser instincts and, at its core, merely plays on much of our socio-economic insecurities. When you are struggling to make enough to earn a living or to get a job, it becomes convenient to blame the Other. Ideas that harp on the supposed supremacy of your group appear to be the easy way out. I am convinced that winning the war on ideas of climate change will have to be won with a similar understanding.

Back to being appointed minister: I thought I 'knew' what needed to be known about climate change and green issues. And now, I would oversee Malaysia's success or failure in this regard.

Where There's (Political) Will, There Shouldn't Be a Won't

Whether we save the planet boils down to *political* will and having a compelling narrative to convince the public of what needs to be done to fight climate change. When fighting a formidable enemy during a war, it is best to build a broad alliance. Similarly, while facing the biggest battle for mankind, we must forge alliances to fight climate change. Progressives tend to think that facts alone can help us win public opinion, when, time and time again, vested interests have succeed simply by telling a better story than progressives. Brexit, Donald Trump, or the extreme racial and religious politics we see succeeding in Malaysia are a testament to this and the fact that it is a political reality, no matter where you are.

Countries and companies across the world have now announced ambitious goals for climate change. Malaysia has committed to achieving net-zero status by 2050. This reminds me of something from my school days. My friends and I would put up big posters with our ambitious exam targets to motivate us. Most of us—including me—would then let the targets be targets without putting in the effort to achieve them. We must not let this happen to our grand targets for curbing climate change. Instead, we should forge a path toward achieving net-zero and fighting climate change. The time to act is now. We have passed the phase of trying; we are now in the phase of doing.

Let us imagine, for a moment, a world where capitalism is in overdrive, where everyone pollutes and wastes resources without control. Our economy will come to a standstill. Our health will deteriorate thanks to bad air and polluted rivers. Our cities will be wastelands. Humans, animals, and plants will perish. All of this will primarily happen due to human greed, arrogance, and foolishness of putting profit and convenience before our mutual responsibilities to each other and the future. After all, we are living in Anthropocene—the age of human-induced climate change.

It Boils Down to Values

The green politics of climate change and environment is sometimes portrayed as a modern concept, having nothing to do with the religious traditions of the world. This perception can come from secular fundamentalists who are sceptical of religion or religious extremists who take a close-minded, literalist view of faith. Among Muslims, some say that climate change is God's punishment to mankind for their immorality or a sign of the end of times. Others take a more political perspective that climate change and environmentalism are Western concepts that weaken the developing (and Muslim) world. Of course, many of the wealthy oil and gas powers happen to be Muslim countries.

In Islam, we are taught that since all creation ultimately belongs to God, mankind is merely the steward of this world. We do not have absolute ownership over it and must care for it before it passes to the next generation. Furthermore, the Quran teaches us that Islam is a mercy not just for believers or even human beings, but for God's creation at large, including animals and plants. More needs to be done to remind Muslims about how central the environment is to the faith. The Quran says:

> Corruption doth appear on land and sea because of (the evil) which men's hands have done, that He may make them taste a part of that which they have done, in order that they may return.[8]

Back in 2015, I stated in a keynote speech at the World Affairs Council in Jordan:

> ... we need to embrace the treasures of Islamic civilisation to provide succour to a world that has realized that contemporary definitions of development have only left us with a broken capitalist system, looming environmental disasters and a

---

[8] The Meaning of the Glorious Koran: An Explanatory Translation (1930) New York: Alfred A Knopf by Marmaduke Pickthall

disillusioned, apathetic electorate. At the heart of these crises is the search for meaning—and on this, we have much to offer.

Image 2: Meeting the former Grand Mufti of Egypt Sheikh Ali Gomaa, an outspoken proponent of an active role for faith in protecting the environment in Cairo

Four years later, I visited Shaykh Abdul Hakim Murad, the British Muslim scholar. Azman Mokhtar, who was taking a sabbatical after his tenure as the managing director at Khazanah Nasional, arranged the visit. Before our group was invited for formal dining at St Catherine's College, the Shaykh brought us to pray at the new Cambridge Central Mosque. This was a privilege, as it was not open to the public yet. It was dubbed as Europe's first eco-mosque with room for 1,000 people. The patron of the project is the famous musician Cat Stevens. We were shown the various green features: recycled rainwater for the gardens, energy harvesting heat pumps that regulate the temperature of the mosque, and large skylights in the main prayer hall to reduce lighting costs, among others. This shows that sustainability is no stranger to Islam.

Interestingly, in Pew's research across twenty-two countries in 2010, the citizens of Türkiye (74 per cent) and Lebanon (71 per cent) ranked second and third in their concern for the environment. The US ranked twentieth (37 per cent) and the only Muslim country ranked lower was Pakistan (22 per cent). Malaysia was not included in the survey.

Similarly, Christianity, too, upholds the principle of human stewardship of this world. Pope Francis in *Laudato Si'* *(Praise Be to You)* focuses on caring for our common home—Planet Earth. The Head of the Catholic Church calls for radical reforms and for developed countries to take responsibility for the environmental and climate crisis. In his follow-up Apostolic Exhortation, *Laudate Deum (Praise God)*, Pope Francis wrote:

> It is often heard also that efforts to mitigate climate change by reducing the use of fossil fuels and developing cleaner energy sources will lead to a reduction in the number of jobs. What is happening is that millions of people are losing their jobs due to different effects of climate change: rising sea levels, droughts and other phenomena affecting the planet have left many people adrift. Conversely, the transition to renewable forms of energy, properly managed, as well as efforts to

adapt to the damage caused by climate change, are capable of generating countless jobs in different sectors. This demands that politicians and business leaders should even now be concerning themselves with it.[9]

Presbyterians have played a pioneering role in conservation. Calvinists are taught to find the glory of God in nature.

In the Chinese tradition, the great Confucian sage Mencius, who died in 289 BC, was already arguing about sustainably harvesting forests. The Songyang Academy was established in the eleventh century on the site of two sacred trees that were already about a thousand years old at that time. Emperor Wu of the Han Dynasty had visited the site to pray there. In 2013, the Confucianists launched a specifically environmental movement as part of their tradition. Proper respect for nature has been a longstanding pillar of Chinese civilization.

The Vedas and the Upanishads teach Hindus that there is no separation between the sacred and nature. The Hindu pantheon of deities, *devas*, expresses itself through natural phenomena: earth, water, fire, rain, and wind. Rivers, too, are represented by devas, and trees and forests are revered as the abode of devas. The concept of karma, too, means Hindus are rooted in the belief of action and consequence beyond our narrow, selfish interests. Therefore, any positive environmental action brings good karma, while bad actions bring about bad karma.

The principles of *ahimsa* (do no harm), *karuna* (compassion), and *metta* (loving-kindness) provide a foundation for the way Buddhists look at the environment. The path to enlightenment is rooted in the concept of the Middle Way or moderation between rigid abstinence and extreme hedonism. Today, the 14th

---

[9] Pope Francis, 'Laudate Deum', *Vatican*, October 4, 2023, https://www.vatican.va/content/francesco/en/apost_exhortations/documents/20231004-laudate-deum.html [Accessed October 7, 2023].

Dalai Lama emphasizes the need for balanced and responsible development. Vietnamese monk Thích Nhất Hạnh has argued that the planet and the self are one and the same.

There is an urgent need to remind religious communities about how central the environment is to faith. But it is also crucial to remind policymakers, scientists, and economists that a moral and ethical vacuum has led us to where we are today. We cannot merely blame science and technology for the excesses of the Industrial Revolution, for at the heart of the problem is the longstanding fundamental issue of greed. American environmental lawyer and former administrator at the UN Sustainable Development Group Gus Speth accurately said:

> I used to think that the top environmental problems were biodiversity loss, ecosystem collapse, and climate change. I thought that thirty years of good science could address these problems. I was wrong. The top environmental problems are selfishness, greed, and apathy, and to deal with these we need a cultural and spiritual transformation. And we lawyers and scientists don't know how to do that.[10]

At the same time, engaging with religious leaders to convey this message is crucial to ensure that the environment and climate change are issues that reach the masses, and do not remain urban, middle class or youth issues alone.

I had the opportunity to deliver a keynote speech at the Faith Pavilion at COP 28 in Dubai. Throughout the conference, speakers representing nine faiths spoke, reflecting on the different religious perspectives as well as common wisdom

---

[10] Speth, Gus, as quoted by Hamid, Zakri Abdul, 'Cultural and Spiritual Transformation Needed', *New Straits Times*, July 24, 2019, https://www.nst.com.my/opinion/columnists/2019/07/507031/cultural-and-spiritual-transformation-needed [Accessed October 31, 2023].

that binds religious communities together as they deal with climate change and environmental degradation. The panellists included Dr Chris Elisara from the World Evangelical Alliance, Chief Dr Doliwura Zakaria from the African Union's Interfaith Dialogue Forum as well as other politicians and academicians.

Climate justice is, hence, an important concept both at the intra- and international level. Industrialization and colonization over the past two centuries have turbocharged the exploitation of resources among developed countries. Now, the developed world is pushing developing countries to adopt higher environmental standards despite the need for the latter to catch up with attaining higher living standards.

Similarly, within every country, Malaysia included, it is the poor who disproportionately bear the consequences of the climate crisis. Simply providing a prescriptive approach, whether from the industrialized nations or richer segments of society, without a just financial structure, will not encourage the adoption of necessary environmental and climate ambitions by those more concerned about immediate economic issues. Yes, following the footsteps of the developed countries in growing their economies is not sustainable, but without proper assistance from those countries—if not in some form of reparation—the developing world will not be able to progress sustainably. At the end of the day, our one and only planet will suffer. Therefore, addressing climate justice is crucial.

In short, Malaysia and the wider world do not have *technological* and *physical* difficulties to achieve our climate change goals. Even *economic* challenges can be addressed. During events with bankers and financiers, I like to repeat: 'We must green the finance and finance the green.' It is not enough for companies to merely focus on avoiding being in the red and to be in the black. They need to be in the green as well.

The economic consequences of the triple planetary crisis are real. It is often necessary to get governments, business, and the public to see the cost of doing nothing. But, by reducing everything to dollars and cents, we still do not get to the core of the issue. We can only see things clearly when we realize that human beings are part of, not separate from, nature. The dignity of man and sanctity of human life is something that is thought of by all religions and traditions. Therefore, nature is priceless and the value of preserving it is paramount. Oscar Wilde famously said, 'A cynic is a man who knows the price of everything and the value of nothing.'

Once we establish that nature is central to the survival of mankind, we can build the political will to make the necessary changes, shape the necessary socio-economic policies to sustain them, and scale the available technology to see them through.

I make no apologies for taking a very personal approach in this introduction. As contradictory as it may sound—it's not about me. But the only way to understand our societies and what is happening to them is to see yourself in their wider context. The sociological imagination, which has been defined as the ability to see the conditions that shape your decisions as well as those of others, is crucial.[11]

This book is an outline of my experiences so far at NRECC and NRES. At the same time, it describes my vision for Malaysia with regards to environment and climate change as well as how we can contribute at a regional and, dare I say, global context. I have been inspired after meeting well-informed, passionate young activists who remind me that what we, the older generations, do today will have an impact that lasts long into the future.

---

[11] National University, 'What is Sociological Imagination', n.d., https://www.nu.edu/blog/what-is-sociological-imagination/ [Accessed November 8, 2023].

Image 3: Speaking at the post-cabinet press conference with Prime Minister Anwar Ibrahim and Transport Minister Anthony Loke.
Photo credit: Faisol Mustafa / Utusan

Prime Minister Anwar Ibrahim captured this when he said:

> We cannot be sustainable, if we look purely in terms of technological change or transformation without care, concern and compassion.[12]

At the end of the day, we owe it to our children and their children to leave them a liveable and sustainable planet. I want them to enjoy the same wildlife and plants that we do today. Ultimately, I want them to have a planet to live on and not inherit a hellish

---

[12] Rahimy Rahim, 'Technological Change in the Nation Must Bring Benefit to All, Says PM', *The Star*, September 27, 2023, https://www.thestar.com.my/news/nation/2023/09/27/technological-change-in-the-nation-must-bring-benefit-to-all-says-pm [Accessed October 19, 2023].

wasteland. I am sure you think the same—that is why you are reading this book. The future generations have more at stake so if, today, we do not do anything, they will not be able to carry on the struggle and will never forgive us because we have one last shot at saving our earth. I hope to show what is at stake for a developing country like Malaysia, what we are doing to meet this challenge, and what the world needs to do if we are to have any future on this planet. In short, I hope to show that there is hope.

On the Shoulders of Giants

'But man is a part of nature, and his war against nature is inevitably a war against himself.'

—Rachel Carson, *Silent Spring*

The NRECC and NRES ministries have a long history. As mentioned, the crucial difference between my former and present ministries is that NRECC includes energy transition and public utilities (electricity, water, and sewage) under its purview, while NRES does not. Both NRECC and NRES are responsible for natural resources, environment, and climate change.

Coincidentally, my father, Nik Ahmad Nik Hassan was the first secretary-general of the energy, telecommunications, and post ministry from 1978 until his retirement in 1988. By the time I entered Cabinet, he was over ninety years old and unwell, but I could now reflect on and relate to the various experiences in the energy sector that he shared with me as I was growing up.

Nevertheless, much has changed. Postal services are no longer a major task for the government, and 'climate change', as we know it, has only emerged in a big way in our discourse along with the term 'global warming' since the 1980s. I had two separate clocking-in sessions at the headquarters of the previous ministries—Energy and Natural Resources—on 5 December 2022 followed by Environment and Water on the next day. I got my team to prepare some briefing notes, but even then, it took some time for me to be properly familiar with my portfolios.

Our Heritage and the Environment

Image 4: I dropped by the Greenpeace's famous Rainbow Warrior III off Port Klang. I later participated in a forum featuring activists including Orang Asli activist and artist Shaq Koyok.

Historically, our heritage has always been in harmony with the environment. The various indigenous Orang Asli ethnicities have been living closely with nature. The Semelai, Temiar, and Cheq Wong, for example, live in or near forests. The Mah Meri and Seletar live near the coast. As they depend on forests and seas for their homes and food, they normally utilize these resources sustainably. They appreciate the value of natural capital as a treasure to be safeguarded, to be handed down to future generations just as they had inherited it.

When Europeans first ventured to India and Southeast Asia in the sixteenth and seventeenth centuries, they were amazed at the local cities. The cities were very cosmopolitan. Malacca, prior to its defeat to the Portuguese, was described by Tomé Pires, apothecary and administrator for the European power, as a crucial entrepot, where eighty-four languages were spoken during the height of its powers. Pires famously described that whoever controls Malacca would basically control the trade into Venice! Yet, the city blended well with its natural surroundings full of luscious trees and shrubs. This contrasted with the dense, barren cities in Europe. Like other traditional houses built by Austronesians, the houses in Malacca were built on stilts. They were well ventilated to deal with the tropical climate.

One European visitor described Aceh, which replaced Malacca as the major power in the Strait of Malacca after the Portuguese conquest, as:

> Imagine a forest of coconut trees, bamboos, pineapples and bananas [. . .] put into this forest an incredible number of houses [. . .] divide the various quarters by meadows and woods; spread throughout these forests as many people as you see in your towns [. . .] you will form a pretty accurate idea of Aceh. Everything is neglected and natural, rustic and a little wild. When one is at anchor one sees not a single vestige or appearance of a city, because the great trees along the shore hide all its houses.[13]

The Dusun community in Sabah follows the *tagal* tradition. It involves prohibition to protect the ecosystem. The community is prohibited from taking freshwater fish out of the river, which conserves the fish population. The river is divided into zones where no fishing is allowed at all, where fishing for sustenance is allowed with the community's agreement, and where fishing is

---

[13] Reid, Anthony, 'The Structure of Cities in Southeast Asia, Fifteenth to Seventeenth Centuries', *Journal of Southeast Asian Studies*, vol. 11, 2. September 1980, p. 241.

allowed for the locals and done under the collective *gotong-royong* concept once a year. The concept, which today is administered by the Sabah state fisheries department, was born from a Dusun myth on the close friendship between man and fish. Fishes such as *pelian* (Malaysian mahseer) benefit from tagal. In the no fishing zone, the villagers have trained the fish to perform fish massage for visitors. The fish nibble at the feet of the visitors, and those suffering from psoriasis (dry and scaly skin) find relief from this therapy. The department has tried to introduce tagal to other communities in Sabah, but since this tradition is not rooted there, it has not caught on.

In front of the Penang High Court, which is over a century old and built in the Palladian style, is a memorial to a Scotsman, James Richardson Logan (1819–1869). A lawyer by profession, he is known to be the person who promoted the term 'Indonesia' among the Malay Archipelago and spoke out for the Asians against the British colonial administrators. A less well known fact is that he was also vocal about the environment. He wrote in 1848:

> It was remarked that the whole of the eastern front of the range [of a mountain in Pinang] has within a few years been denuded of its forest [. . .] Climate concerns the whole community and its protection from injury is one of the duties of Government [. . .]
>
> The great extent to which the plain of the mainland of Pinang has been shorn of its forest would of itself produce an urgent necessity for a stop being at once put to a war with nature, which must entail severe calamities on the future [. . .]
>
> We are informed that the destruction of jungles on the mountains of Pinang has been allowed to proceed unchecked for the last two years.
>
> If any of the residents will bring it to the notice of the Governor we are sure from our knowledge of his opinions, with respect to the necessity of preserving hill jungle, that he

will not only make an order on the subject, but what is essential, provide means for carrying it into effect.[14]

The Emergence of the Ministry

After I was appointed minister of NRECC, I scheduled to have lunch with my father's fellow student at the National University of Singapore, Musa Hitam. He was deputy prime minister from 1981 to 1986. At eight-nine years old, he was still his witty and sharp self. Other than sharing stories about his time with my father, both as a student and later during their parallel careers in public service and politics, he shared his experience as minister of primary industries, which oversaw the timber industry. He recalled his experience standing up to the state governments that wanted to continuously cut down forests for quick money.

As the minister for primary industries, Musa accurately observed the pride Malaysia had then as a 'small country yet being the largest exporter of tropical timber' was misplaced. This, he remarked as a cabinet minister, meant that we were proud of 'raping' our own forests!

The first natural resources ministry existed even in Malaya's first cabinet, prior to independence. It was led by Ismail Abdul Rahman who would later end up as Malaysia's deputy prime minister. But, an analysis of parliamentary debates on forest reserves shows that, in the early days, the topic was referred to when discussing the issue of the timber industry rather than in the context of conservation. This started to change in 1979, when environmental elements and their preservation became a consideration.

In terms of Cabinet governance, Malaysia's environmental portfolio emerged with the passing of the Environmental Quality Act 1974. Prior to that, there had only been ad hoc handling of

---

[14] 'No Water Crisis in Penang, as Our Forefathers Protected Our Hills', *The Edge*, April 3, 2014, https://theedgemalaysia.com/article/nutmeg-vista-no-water-crisis-penang-our-forefathers-protected-our-hills [Accessed November 9, 2023].

environmental issues that were often stuck due conflict between Federal and state legislation. Some—such as the Water Act 1920, the Federated Malay States Forest Enactment 1934, and the Merchant Shipping Ordinance 1952—had roots in the colonial era. After Independence, the Land Conservation Act 1960, the Fisheries Act 1963, and the Factories and Machinery Act 1967 were among the related legislation passed.[15] Meanwhile, Malaysia's first environmental NGO was formed prior to the Second World War, in 1940, the MNS).

However, these laws were insufficient to keep up with Malaysia's rapid development. Thus, the Environmental Quality Act 1974 streamlined all the various legislation and allowed the Federal Government the power to act without relying on state legislation. The DOE was established under the 1974 legislation, placed under the new Ministry of Local Government and Environment. This was the first time the term 'environment' appeared in the Malaysian Cabinet. In fact, Malaysia was pretty much at par with the developed world at that point.

Just to cite some comparisons, in 1968, Sweden proposed for the UN to have its first environmental summit. The world's first minister of environment was UK's Peter Walker, who was appointed as secretary of state for environment in 1970. The Department of Environment was one of Prime Minister Edward Heath's two super ministries where three former ministries: public building and works, transport, as well as housing and local government were merged. In the US, the EPA was established in the same year.

One of the major figures that influenced the emergence of the environmental movement as well as the establishment of the

---

[15] Maizatun, Mustafa, 'The Environmental Quality Act 1974: A Significant Legal Instrument for Implementing Environmental Policy Directives of Malaysia', *IIUMLJ*, vol. 19, 1. June 2012, pp. 11–12 https://journals.iium.edu.my/iiumlj/index.php/iiumlj/article/view/1/1 [Accessed June 19, 2023].

EPA was marine biologist and conservationist, Rachel Carson. Her book, *Silent Spring*, focused on the impact of unregulated use of pesticides on the environment.[16] This led to her being attacked by the chemical industry,[17] attracting the attention of President John F. Kennedy, and made concern for the ecology a global movement.

During the same time, environmentalism became a more enduring international movement, partly as a result of the counterculture movement. The first green parties were established in Australia, New Zealand, the UK, and Switzerland. The UN organized the Conference on Human Environment in Stockholm in 1972. Domestically, environmental NGOs such as CAP and EPSM were formed in the 1970s. In 1972, just over a decade after its establishment, WWF Malaysia started its operations. Indeed, despite the country's advanced legislation, the trend in civil society was arguably rooted in a recognition of the failure of the government and the market.

As the state-led capitalism under Prime Minister Dr Mahathir Mohamad coupled with the global move towards privatization became more dominant in the 1980s, environmental civil society began to provide stronger alternative visions for Malaysia. Mahathir responded with the strongest condemnation he could think of, accusing activists of being 'crypto-socialists'.[18] But, by 2008, when the opposition PR coalition made huge gains, the role of civil society became more prominent and they began to enjoy

---

[16] Carson, Rachel, *Silent Spring*. Boston: Houghton Mifflin Company, 1962.

[17] Dreier, Peter. (2012). How Rachel Carson and Michael Harrington Changed the World. Contexts, 11(2), 40-46. https://doi.org/10.1177/1536504212446459 [Accessed August 28, 2024].

[18] Cooke, Fadzilah Majid and Adnan A. Hezri, 'Environmental Activism in Malaysia: Struggling for Justice from Indigenous Lands to Parliamentary Seats', in *Environmental Movements and Politics of the Asian Anthropocene*, Jobin, Paul, Ho Ming-sho, and Hsin-Huang, Michael Hsiao (eds.), Singapore: ISEAS, 2021, pp. 203–231.

a more positive relationship with the authorities. Many of my elected colleagues were from civil society after all. Environmental politics became mainstream.

For so long, the challenge of environment and climate administration in Malaysia has been fragmented governance across different ministries and between the Federal and state governments, as well as the special autonomous status of the regions of Sabah and Sarawak. The National Policy on Biodiversity falls under the purview of NRES. But safeguarding fisheries and marine protected areas falls under the purview of the Ministry of Agriculture and Food Securities. The Ministry of Transport regulates maritime industries. Forests, the main habitat of land-based wildlife, fall under the jurisdiction of state governments. However, in Peninsular Malaysia, the Forest Department of Peninsular Malaysia (under the NRES) manages the forests together with the respective state forestry departments. Sabah and Sarawak have their own forestry departments that are separate from the Peninsula department. The management of sustainable forestry for the extraction of timber falls under the Ministry of Plantation and Commodities. In 2021, it was estimated that there were thirty-two legislations regarding biodiversity protection in Peninsular Malaysia (both Federal and state), seven in Sabah, and eight in Sarawak.[19]

A fragmented approach will not do as the country embarks on its ambitious and necessary journey towards net-zero, and protecting the health, livelihood, and future of its citizens. Our priority is to reduce carbon emissions. A business-as-usual approach is no longer relevant. But even with the best efforts, we can only do so much. Hence, carbon sequestration is crucial. It is the best nature-based solution to climate change—in forests,

---

[19] Reef Check Malaysia, 'Political Will in Short Supply', April 28, 2021, https://www.reefcheck.org.my/blog/political-will-in-short-supply [Accessed October 25, 2023].

oceans, coasts, cities, and farmlands through our forests, trees, peats, and oceans—that doesn't entirely dismiss the role for CCS.

A major source of carbon emissions is energy, and that needs to be mitigated. However, mitigation, too, has its limits, and climate change adaptation and resilience to the current state of the environment is crucial. Here, water supply, floods, and coastal erosion are all pertinent issues. The task of getting all the different moving parts to work together for our net-zero objective is crucial and urgent.

Energy Transition

'The future is green energy, sustainability, renewable energy.'

—Arnold Schwarzenegger, former actor and governor of California 2003–2011

Electricity has played a key role in delivering better living standards throughout the world. High electricity consumption and high human development are generally correlated. The advantage of this form of energy is that it is easy to transport and transform into different forms of energy. Electricity as a form of energy is not polluting at the end-use, for example, using electric lighting and heating as opposed to gas lighting or heating. This gives it an advantage compared to other energy sources. Better lighting allows us to be more productive at night and increases safety. The advent of air-conditioning has not only allowed us to work in the stifling tropical heat but has also reduced the risk from heatwaves. Many industrial and medical devices are only function due to electricity. However, much of electricity is generated from fossil fuels such as coal, oil, and gas, which are polluting and emit greenhouse gases in the atmosphere. Thus, transitioning to cleaner and more sustainable electricity is crucial in achieving our net-zero objectives. At the same time, diversifying sources of electricity generation is also crucial for energy security.

I was a fourteen-year-old student at MCKK and undergoing my Boy Scouts' camping programme in Sayong, just across the

Perak river. Like any teenager, I was excited to be out in the open, setting up tents and meeting friends from different schools in the district. We paid 30 sen each for the boat ride to cross the mighty river. When we were on the Ipoh–Kuala Kangsar Road at night, suddenly everything, from the streetlamps to the houses, went pitch dark—the lights had gone out. This was 1996, and later, we would find out that this was a major blackout that affected many parts of Peninsula Malaysia. The FA Cup final between Kedah and Sarawak was supposed to take place that night, and tens of thousands of fans were disappointed when the match was cancelled. Businesses reported losses in the hundreds of millions of ringgit at that time. Traffic went into gridlock.

This was four years after the country had suffered another major blackout. Nine states faced electricity cuts in 1992 after lightning struck a transmission line in Terengganu on the East Coast of Peninsular Malaysia, causing a rolling failure in the grid. This led to twelve power plants on the West Coast losing power. It took up to two days for power to be fully restored.

Malaysia was undergoing a process of massive economic growth and it was feared that the blackouts would dent investors' confidence. These problems with electricity supply became an excuse for the government to introduce IPPs. These are power plants not owned by the national utility, TNB. Prior to that, TNB and its predecessor, the National Electricity Board, monopolized a range of activities from generation and transmission to distribution of electricity in the Peninsula. The IPPs were allowed to build and operate power plants, which, in turn, would sell electricity to TNB for transmission and distribution through a power purchase agreement. From a 23 per cent reserve margin capacity in 1991, the country went up to almost 50 per cent five years later.[20] These high reserves impacted the costs that the

---

[20] Smith, Thomas B., 'Privatising Electric Power in Malaysia and Thailand: Politics and Infrastructure Development Policy', *Public Admin. Dev.*, vol. 23, 2003, 273-283, p. 277.

government bore and the tariffs charged to the consumers. At the same time, both TNB power plants and IPPs receive gas subsidies from Petronas (Petroliam Nasional Berhad).

Back in 1978, when my father assumed the position as the secretary-general for the new Ministry of Energy, Telecommunications, and Post, the challenges facing the Malaysian energy sector were developmental. My father worked under Minister Leo Moggie Irok, who had held that portfolio until 1989. Leo Moggie was reassigned the energy portfolio in 1995 until 2004.

The World Energy Council introduced the concept of the energy *trilemma* that is faced by policymakers to balance affordability, security, and sustainability. Universal access to energy that is within reach of all segments of society addresses the issue of energy affordability. The ability to source fuel and produce electricity at present and in the future through a reliable grid addresses the issue of security. When I held the position of NRECC minister from 2022–2023 and was in charge of electricity, I liked to joke that in my father's time there had only been the *dilemma* of affordability and security. There was little if any talk of green or renewable energy and sustainability. After Malaya became independent in 1957 and Malaysia was formed in 1963, the government was understandably focused on providing reliable electricity across the country, particularly bringing the benefits of electrification to the rural areas. Today, sustainability is emerging as an important component on its own with the looming climate crisis.

The energy mix has also changed due to economic as well as environmental factors. From the 1960s, most of the electricity generation came from oil. This changed following the oil price shocks of the 1970s. In 1973, the world suffered an oil crisis, known as the first oil shock. OPEC agreed to increase the price of oil. The price of a barrel of crude oil in the US increased five-fold. It was US$2.75 per barrel in January 1973 and continuously shot up until March 1974 when the price was US$12. This was then compounded by the 1979 energy crisis caused by a disruption of Iranian oil production, following the Iranian Revolution, and then both Iranian and Iraqi oil production following the invasion of Iran by Iraq. The price of a barrel of crude oil was US$35 in 1980. This became known as the second oil shock.

During these oil price shocks, there were attempts at building an early form of renewable green economy at the global level. But these efforts were dwarfed by the 'grey' technologies: the widespread availability of fossil fuels meant investing in fossil fuel efficiency was the main path taken by many corporates. Cars became more efficient. The perverse effect was that demand for fossil-fuel-powered vehicles increased, as one could go further using the same amount of petrol. Petrol sales increased. Better housing insulation in temperate housing meant that people became more accustomed to warmer temperatures indoors and dressed lesser in colder countries.

By the early 1980s, hydroelectricity and then a small amount of natural gas was being used to generate electricity. Malaysia had a policy to address depletion of oil reserves introduced in 1980. As a result, our use of oil for electricity generation fell and gas increased, as it is found abundantly in the country. Many IPPs initially decided to build gas-powered plants due to the relative ease of setting them up. However, gas was afterwards included under the National Depletion Policy 1980. Coal became a significant part of the energy mix starting in the 1990s.

Renewable energy was included as the country's 'fifth fuel' at the turn of the new millennium, alongside oil, gas, coal, and hydroelectricity. This was following the 1997 Kyoto Protocol of the UNFCCC that committed countries to reduce greenhouse gas emissions. Nevertheless, renewables did not take off immediately at that time, while ironically, the use of coal grew tremendously.

However, things are different today. Green alternatives are cheaper and more readily available while fossil fuels are more volatile and less reliable.

Low-Cost, Export-Oriented Economic Model

In the 1980s, the focus of Malaysia's economic policy shifted to provide a low-cost, export-oriented environment to attract foreign investments. With low costs—suppressed wages, cheap utilities, and affordable land, as well as minimal taxes—investments poured into the country. This strategy is not unusual. We followed the model of many Asian economies—like Japan, South Korea, Taiwan, and Singapore—to kick-start our development with industrialization, just as developed economies were pushing dirty, heavy industries out.

Once, at a meeting with then Prime Minister Dr Mahathir Mohamad, which my father attended, the Prime Minister was complaining about increasing electricity tariffs as the price of crude oil surged. At that time, diesel was one of the main fuels for power generation. Told by the people at the National Electricity Board (LLN) (as TNB was known prior to privatization) that this was due to how the tariff was structured, Mahathir wanted something to be done. My father shot back that this would mean the government picking up the tab, and the Prime Minister said he was prepared to do so.[21]

Thus, electricity tariffs were kept low for all consumers in the form of a blanket subsidy for a long time. This was central to our low-cost, export oriented industrial economy. Wages and corporate taxes were kept low. Due to low wages, the government had to address the political issue of keeping costs of living low,

---

[21] Nik Nazmi Nik Ahmad, *In the Public Service: The Life of My Father, Nik Ahmad.* Kuala Lumpur: Pusat Sepakat, 2018, p. 118.

and one simple method was to have blanket subsidies to ensure lower electricity tariffs. The economy suffered from a structural problem of a historically low share of labour income in the economy, whereas capital income—profits for shareholders—were high.

Therefore, blanket subsidies are a band-aid solution. Now, we need a more holistic solution to the issue. In fact, perversely, it is the rich who benefit most from blanket subsidies. The government was and is suffering from a low revenue base, with less to spend on public services.

Power Privatization

Inspired by Ronald Reagan in the US and Margaret Thatcher in the UK, Mahathir moved to privatize utilities, including the electricity sector. Officially, the privatization policy was announced in Malaysia a year later. The justification given was that the private sector is more efficient, and the government should only focus on essential sectors. The logic was that the employees would be paid better; the privatized utilities could look for more outside funding to invest in infrastructure, while the government would benefit from a more streamlined public sector. At a time of rapid change, privatization was also seen as a means for utilities to catch up with the latest technology.[22]

In the government document Privatization Guideline 1985, the five objectives of privatization were:

1. reducing the financial and administrative burden of the government;
2. encouraging competition, efficiency, and productivity;
3. increasing the pace of growth through privatization;

---

[22] Ibid., p. 115.

4. reducing the size and participation of the public sector in the economy;
5. providing opportunities to achieve the goals of the NEP.

Chile was among the early models for electricity sector deregulation in 1982. In Malaysia, overall, TNB has been successful in delivering reliable energy, although the creation of IPPs has been controversial. We have seen the failure of water and waste water privatization in Malaysia, leading to the renationalization of the utilities. Political interference and opaque privatization processes have hindered privatization achieving the objectives outlined above.

The privatization of TNB was followed by the creation of IPPs mentioned at the beginning of the chapter. In 1993, the Federal Government revived the Bakun Project in the Borneo state, Sarawak. There was an old proposal—dating back to the 1960s—to build a hydroelectricity dam over the Balui river that would generate 2,400 megawatts of electricity, way more than what Sarawak needed. The idea had been to build a 670-kilometre submarine cable to Peninsular Malaysia. However, the increasing reserve margins and cheaper alternative of utilizing natural gas had meant that Bakun was not economical at that time. In the revised version of the project, it was privatized directly to a timber company, Ekran Berhad, which focused on the 'low hanging fruit' of the business end of the deal: clearing virgin jungles and extracting timber from an area the size of Singapore that would be underwater once the dam was completed. The project stopped due to the 1997 Asian financial crisis. The Federal Government had to bail out the project.

Transition to Targeted Subsidies

The ICPT mechanism was first developed in January 2015. Essentially, the implementation of this mechanism means that electricity tariffs more transparently reflect changing fuel prices and other generation-related costs. Every six months, TNB calculates the tariff for its consumers based on the cost of coal and gas over the last six months. It is estimated that 65 per cent of the cost of electricity in Malaysia is due to fuel prices. The idea of the ICPT is to allow the fluctuating price of fuel to be passed through to the consumers.

Less than two weeks into my job as NRECC minister, after my papers were approved by the Cabinet, I was tasked with announcing the ICPT for the January to June 2023 period. It was a demanding and daunting task, involving a very technical subject for a rookie minister. I spent time poring over the statement, the explanatory notes, and the different numbers. I initially planned to take the safe approach of releasing a press statement via a PDF file. But Rafizi Ramli, the economic minister who has been a mentor to me in politics, advised me to take charge of the issue by holding a press conference and fielding questions from the media. Since this was a political decision, he suggested that I should own it: 'That way, you will truly master the brief and not be merely reliant on the civil servants.'

The government decided to keep the 2 sen per kilowatt hour rebate for domestic tariffs while increasing the surcharge for non-domestic high- and medium-voltage customers from 3.7 sen to 20 sen. This was a massive increase on the surcharge of 410 per cent, but even then, the government was subsidising 7 sen. Many businesses complained, and I met to engage with some of their representatives. They were calling for a moratorium. I explained to them that while the increase was steep, they had enjoyed huge subsidies before, and it was not sustainable for the government to continuing doing so. There were more deserving electricity customers who deserved help from the government. The low-cost model was relevant for kick-starting our development, but to be competitive, Malaysia had to transition to a high-value economy. A few months later, the complaints subsided somewhat.

At the same time, there were many reports of affected businesses installing solar panels and energy efficiency measures to reduce their energy bills.[23] One of the hindrances to the adoption of renewable energy in Malaysia for so long were the blanket subsidies. We were comforted to see the policy changes having an immediate impact. There had been incentives to install solar panels previously, but without addressing subsidies for electricity generated using fossil fuels, the policy would not go very far. It was as if one hand was giving money to make renewable energy cheaper while another hand was giving money to fossil-fuels-derived electricity cheaper.

When the July to December 2023 ICPT cycle came, we took the lead from Prime Minister Anwar Ibrahim's stand that eventually subsidies would be removed from the richest households. Only a small number of domestic accounts were affected. The government increased the tariff for domestic households consuming more than 1,500-kilowatt hour per month. This accounted for approximately the top 1 per cent consumers (and those with electricity bills of more than RM740 a month) because they would have a 10 sen per kilowatt hour surcharge as opposed to the existing two sen per kilowatt hour rebate for all domestic consumers. These consumers still received subsidies, albeit at a reduced rate.

It is important to note that the tariff adjustments were done while waiting for the centralized household database, PADU, which would provide detailed socio-economic data of households in Malaysia, to be ready. Our government planned for the database to be ready by 2024. Thus, we had to rely on consumption data,

---

[23] Gan, Khoo Sek, 'SMEs Turn to Solar Power', *The Star*, April 9, 2023, https://www.thestar.com.my/news/nation/2023/04/09/smes-turn-to-solar-power [Accessed June 25, 2023]; 'TNB Unit to Install Solar PV Facilities at Aeon Malls', *The Star*, April 13, 2023, https://www.thestar.com.my/business/business-news/2023/04/13/tnb-unit-to-install-solar-pv-facilities-at-aeon-malls [Accessed June 25, 2023].

which is a rough indicator of household income, and not the more accurate household income. When the database—which will look at income, benefits, and number of dependents in a household—is ready, it will allow subsidies to be targeted more accurately based on household income.

Even then, many ministers and politicians were nervous about the decision, especially as state elections were around the corner. Some accused us of engaging in the rhetoric of class-warfare. In fact, some civil servants were jittery as well. I argued that we need to keep pursuing the transition to targeted subsidies to help us make difficult decisions in the future. If not, we would just be kicking the can down the road and would face even more drastic hikes in the future. The Prime Minister summed it well by saying that when we donate, we give to those who need it, not the wealthy. Subsidies are essentially the same and should be targeted towards the working class and maybe part of the squeezed middle class in Malaysia.

In response to the announcement, *The Edge* commended that this was 'another step in the right direction'.[24] As a result of the transition to targeted subsidies in January and July 2023, we saved RM4.6 billion. As an indication, this money is enough to build fifty-seven new schools, fifty new district hospitals, or 180 new health clinics. That amount is also sufficient to employ 14,000 new doctors for five years.

Nobel-Prize-winning economists Abhijit V. Banerjee and Esther Duflo wrote that energy consumption is like addiction, and the best response to change behaviour for the future is high taxes. In Malaysia, as we subsidize electricity, removing blanket

---

[24] 'A Step in the Right Direction', *The Edge*, June 26, 2023, https://theedgemalaysia.com/node/672499 [Accessed June 30, 2023].

subsidies is a start.[25] This is the structural change that we need to transition from a growth-at-all-cost model to one that is fair for everyone and sustainable for the planet.

Energy Efficiency

Being a politician, I know that for many, building massive solar farms, biomass or, in the future, hydrogen power plants is always sexier than focusing on switching to light-emitting diode (LED) bulbs and energy-efficient appliances or insulating a building to reduce the use of cooling energy. The optics of a politician cutting the ribbon of a shining new large-scale solar plant or posing on an excavator to break ground for a biomass power plant construction is always what the communications advisor orders. It looks good on Instagram and attracts a lot of engagement. Using a football analogy, energy efficiency is like Liverpool investing in an academy player like Trent Alexander-Arnold as opposed to a massive solar plant, which is like splurging on players from outside like Virgil van Dijk, who cost GBP75 million in 2018 and was then the world's most expensive defender. In short, from a political and public relations perspective, renewable energy trumps energy efficiency.

But the reduction of energy consumption is also key for us to cut greenhouse gas emissions. In fact, energy efficiency is deemed as the first fuel for energy transition because, often, it is the quickest and cheapest way to mitigate carbon emission. Capitalizing on this low hanging fruit is important in our journey to net-zero. Even with the advent of industrialization since the nineteenth century, it is only since the 1960s that our consumption of fossil fuels has increased substantially to fulfil

---

[25] Banerjee, Abhijit V and Esther Duflo, *Good Economics for Hard Times*. London: Penguin, 2020, p. 220.

our increased energy demand. We are deluding ourselves if we think that decarbonization can be achieved via renewable energy alone, especially with the demands on infrastructure. We cannot ignore reducing demand for electricity.

Electrical appliances from refrigerators to aircons and televisions to even electric bulbs are more efficient at using electricity today than in the past (I, for one, am from the generation that used the huge televisions and computer monitors when I was growing up!). Higher energy prices are also pushing people to be more careful about using electricity compared to the past.

Since peaking in 2005, the UK's electricity generation per person has fallen steadily, despite a population growth of almost 12 per cent. This is partly due to energy efficiency, use of private rooftop solar panels that are not measured in the grid, as well as deindustrialization. At the same time, household and business consumers have been *paid* by the grid to use electricity at non-peak times. This has saved electricity use during high demand hours and prevented blackouts from happening.

Among the developed economies, Japan suffered more than most during the 1970s' oil price shocks due to its lack of indigenous resources and dependence on imported oil. The price of goods in the country increased sharply. Electricity tariffs surged. The country began to focus more on generating electricity from nuclear power, having had to convince a generally hostile public about it, particularly the survivors of the Hiroshima and Nagasaki bombings. Taxes were levied on petroleum products. The use of alternatives—natural gas (and unfortunately coal)—began to grow. Most of the increases in oil prices were passed on to the consumer. Japan passed a law on rationalization and conservation of energy in 1979. Resources were invested in energy efficiency and use of alternative energy, including the Moonlight Project that tested a battery storage system. This allowed excess energy generated at night to be stored in batteries and then dispatched during the day when demand was higher, particularly due to the use of air conditioning. After facing several challenges, this finally led to a viable battery project with the increased use of renewable energy in the 2000s.

Malaysia had been working on the EECA for quite some time. Since the turn of the century, Malaysia is using more energy to produce the same level of economic growth. In short, our energy intensity is increasing. Energy efficiency is important to ensure we get more with less resources. In January 2018, the Cabinet took note of the need for an efficiency framework to manage energy demand. In June 2019 and November 2020, the Cabinet agreed on the drafting of the EECA. It would only apply to large energy consumers—those who spend over RM80,000 a month on natural gas bills or RM200,000 a month on electricity bills. This would only affect 2,000 installations, equivalent to 0.1 per cent of total installations, but would cover 70 per cent of the biggest energy consumers in the country. For these energy guzzlers, this would be an investment to actually lower their energy bills in the long run.

Finally, five prime ministers and five energy ministers since it was first proposed, in December 2023, we managed to pass the EECA in Parliament. After the bill was passed in the lower house, I was greeted by the happy faces of NRECC civil servants and officers of the ST that had been working hard over the years to discuss, draft, and eventually push through this piece of legislation.

The bottom line is, as is discussed in the following sections on the importance of grid, the cost of upgrading the grid to accommodate the ambitious renewal energy supply is big. There is a limit to how many resources can be invested in upgrading the grid: money, material, and human capital to keep pace with variable renewables. We must accept that there are limits—energy efficiency and conservation play a key role in overcoming them.

Renewable Energy

Most of the energy we have on earth comes from the sun—both renewable energy and fossil fuels. Solar energy obviously comes from the sun. It is generated in real time based on sunlight that reaches earth. The heating of the atmosphere and the ground creates wind, which drives wind turbines. Water evaporated by the sun turns into clouds, which then create rain and produce streams and rivers that power hydroelectricity. Plants store energy from the sun through photosynthesis and can be used for biomass. But biomass that is buried under the earth for millions of years, subjected to pressure and decay, become fossil fuels such as petroleum, gas, and coal.

Less than a year after the formation of the Unity Government, we came up with the NETR, which charts the plan to radically grow the share of renewables in our energy mix. It was launched at the end of August 2023 by Prime Minister Anwar Ibrahim, in the presence of Economic Minister Rafizi Ramli and I. Before that, we announced a 'no new coal plant' policy to push the authorities to introduce low-emission and renewable energy alternatives to power the nation.

During COP 28, France launched the Coal Transition Accelerator with the EC, US, UK, Senegal, Canada, Vietnam, and Malaysia. French President Emmanuel Macron invited me to share Malaysia's views. US Climate Envoy John Kerry, Vietnam Prime Minister Pham Minh Chinh, and EC President Ursula von der Leyen were among those attending. I mentioned the challenge for many Asian countries is our coal plants being younger than those in the West. This means any move for early retirement is very expensive. Nevertheless, I said we are looking to request information on carbon reduction and possible early phase out of coal—through early retirement, mothballing, co-firing, and brown to green swaps.

Image 5: Malaysia was among the ten countries and agencies that took part in the launch of the Coal Transition Accelerator at COP28, led by France. Here, French President Emmanuel Macron listened to my point that developing Asian countries need more assistance to embark on this mission.

Mothballing involves cutting down coal power plants' generation, reducing carbon emissions while allowing the plants to be ramped up when necessary to meet power shortages. This is cheaper and faster than starting inactive coal plants. Germany and China adopted this measure, only switching the plants on when needed. Co-firing means adding other fuel, such as ammonia and biomass, to generate electricity with lesser carbon emissions. This has been implemented on a small scale in Malaysia. Brown to green swaps entail some form of early retirement of coal, wherein the IPP is compensated with licenses to produce renewable energy.

We have seen what has been done in other countries. In 2008, when the UK Climate Change Act came into place, about 80 per cent of the energy generated there came from fossil fuels. In August 2023, 47 per cent of electricity came from zero-carbon sources. Since then, the UK has embarked on the fastest transition into renewable and low-carbon energy for a major global economy. It has been highly reliant on wind and solar power. The level of renewable energy in the mix has increased at a much faster rate than predicted.

Image 6: I joined the rest of the NRECC officials, along with TNB and Vantage RE, to visit the majestic Blyth Offshore windfarm at the Northeastern England, UK.

As NRECC minister, I had the opportunity to visit the Blyth Offshore Demonstrator in the UK, a joint venture between TNB subsidiary, Vantage RE, and EDF. Located in the north-east of England, it consists of five offshore wind turbines, each more than 100 metres in height from sea to the rotor. If measured from the tip of blade at its maximum height along with the height of the rotor to the sea level, it is more than 190 metres tall. This is more than twice the length of the wingspan of Airbus A380 and twice the height of the Big Ben! We took the boat to visit the turbines and were shown around by British and Malaysian engineers who maintain it. I noted on my Instagram, you never realize how big and majestic wind turbines are until you are actually up close. It was truly breathtaking. I was slightly disappointed that we were not allowed to go up the massive wind turbines!

Unfortunately, although TNB has interests in this project, with the technology available today, the light wind along the equator makes wind energy not viable in Malaysia.

Uruguay is a developing country that, in terms of GDP per capita or ranking in the Human Development Index, is like Malaysia. Like Japan and unlike Malaysia, it depends heavily on imported oil. Thus, the oil price hike of the 2000s impacted the South American country negatively. The government was forced to import energy from neighbours at more expensive prices. Blackouts became common. Forced to look for a solution, the country turned to Ramón Méndez Galain. A physicist by training, people thought that he would see nuclear power as an option. Instead, Galain opted for wind power, as the country is blessed with ample wind as well as enormous tracts of agricultural land. He was given the chance to be the country's energy minister with one condition—he must convince all the political parties of his energy plan.

Within ten years, around fifty wind farms were set up across Uruguay, supported by hydroelectricity, solar, and biomass.

A software was designed to predict sun and wind patterns a week ahead to ensure that water levels at dams are managed properly. Cattle farms dotting the country began to look at leasing land to wind turbines to supplement their income. Galain attempted to convince politicians and policymakers that even if they are not swayed by the science of climate change, the economics of it should be persuasive enough. Cutting dependence on imported, fluctuating commodities for energy would reduce costs, and the transition would also create 50,000 new jobs. The same argument is now being made throughout the world with the increased price in fossil fuels because of the invasion of Ukraine by Russia. Today, 95–98 per cent of the electricity produced in Uruguay comes from renewables, depending on the weather. When there is an energy surplus, it is exported to neighbouring Brazil.

Closer to home, we can learn from Vietnam. In 2022, 26.4 per cent of electricity in the country was coming from renewable energy (excluding hydroelectricity—solar, wind, and biomass. In 2017, the government issued a feed-in-tariff mechanism for solar power, where its producers would be given twenty-year contracts to sell electricity. At the end of 2018, there was only 86 megawatts of solar photovoltaic capacity. Within six months, this increased to 4,450 megawatts as companies rushed in before the end of the feed-in-tariff period. In 2020, it reached 16,500 megawatts. Vietnam overtook Thailand in terms of installed renewable energy capacity.

Biomass is residue from dead organic material and can be burned to produce energy. Malaysia has been utilizing biomass from oil palms as well as other wood pellets. Every year, we process over 100 million fresh fruit bunches from the crop. The waste—the empty fruit bunch, palm kernel shell, and the remaining pulp—is often left to rot or be incinerated, which contributes to carbon emissions. But they can be made into pellets for biomass plants to generate electricity. In fact, many palm oil mills have biomass power plants. The challenge is the location and relatively small amount of electricity generated, which makes economical connections from the grid to these plants a challenge. The fact is, today, Malaysia is the second largest wood pellet and palm kernel shell exporter in the Asia–Pacific, sending our products to Japan, Korea, and other destinations with developed, large-scale biomass power plants. When we look at developing the renewable energy sector, it is always crucial to draw global lessons and leverage our local advantages to transition.

Solar Energy

Generating electricity from solar is not a new concept. The world's first solar thermal station was built in Maadi, Egypt, in 1913 by American inventor Frank Shuman. Sunlight is concentrated to produce steam, which generates electricity. Cheap coal and petroleum, however, have hindered the use of solar energy. It was only with the advent of the space age, during which there was a need to find a power source for satellites, that solar panels started to gain popularity.

Solar power is the most developed renewable energy source in Malaysia. In 2008, as a fresh (and scrawny) back bench Selangor state legislator, I took part in a debate on the state's developmental plans going into 2025. I was one of the group of young legislators to have won in that year's watershed elections. Referring to a large investment by a major solar panel manufacturer, Q-Cells, worth RM5 billion in Selangor, I argued that it should be treated as an important shift for Malaysia, both economically and environmentally:

> Thank God we have been blessed with a country rich with resources, including plentiful petroleum [. . .] However, these resources will not be there forever [. . .] Therefore, if we do not change our policies where we do not reinvest our God-

> given resources for the long-term benefit of our country, we will be like other petroleum producers, where these riches lull the government, and the government does not productively develop the economy. This is the resource-curse.
>
> Thomas Friedman, the author of *Hot, Flat and Crowded* spelled out the green energy revolution as a main trend to generate economic growth in the 21st century. Therefore, if Selangor is a pioneer to expand the renewable energy industry, we will be one of the pioneers to the most important industry in the 21st century.
>
> Although the pressure for renewable energy may have waned when the price for a barrel of crude oil dropped from US$150 to US$80, but we need to look at the long-term to invest in this sector [. . .] as our petrol resources will not be there forever.[26]

In fact, in the 1980s, the government introduced solar panels for rural electrification. This allowed communities far from the grid to access electricity. Not only is it cleaner compared to diesel-powered generators, but solar power also does not require fuel that is expensive and that needs to be transported to the *kampungs*.

By the time I assumed the position as NRECC minister in 2022, renewables accounted for a quarter of installed electricity capacity of Malaysia. This is in line with the target set in 2021 to achieve 31 per cent renewable capacity in 2025 and 40 per cent in 2035.[27] However this was *capacity*, as opposed to *generation* mix. To ensure we meet our net-zero objectives, we need to look at

---

[26] 'Selangor State Legislative Assembly Debate', October 29, 2008, https://dewan.selangor.gov.my/wp-content/uploads/2019/02/07.HANSARD.291008.pdf [Accessed November 1, 2023].

[27] SEDA, 'Malaysia Renewable Energy Roadmap', 2021, p. 55, https://www.seda.gov.my/reportal/wp-content/uploads/2022/03/MyRER_webVer3.pdf [Accessed June 22, 2023].

the latter. According to ST, renewables made up 16.4 per cent of the energy mix in 2022. Via NETR, we came up with a more ambitious goal, to have 70 per cent capacity by 2050—the earliest year for Malaysia to achieve our net-zero objectives. This means renewable energy capacity needs to reach 55 gigawatts compared to six gigawatts today. A gigawatt is a billion watts. The lightbulb in our homes are typically 100 watts at most. Thus, we need enough renewable energy to power 20 million lightbulbs every year. RM637 billion ringgit is estimated to be needed for Malaysia to make this leap.[28]

Image 7: Prime Minister Anwar Ibrahim sharing a joke with me at the launch of the NETR. In the background is Economic Minister Rafizi Ramli. Rafizi—the person who recruited me into politics—and I worked closely together to come up with NETR, as we pushed a new, ambitious 70 per cent renewable energy capacity by 2050 for Malaysia.

---

[28] IRENA, 'Malaysia Energy Transition Outlook', 2023, p. 114, https://www.irena.org/Publications/2023/Mar/Malaysia-energy-transition-outlook [Accessed August 9, 2023].

In the interim, Malaysia is committed to an unconditional 45 per cent reduction in the intensity of greenhouse gas emissions per unit of GDP by 2030, benchmarked against the 2005 level. NDCs are climate actions that countries pledge to undertake following the Paris Agreement to reduce greenhouse gas emissions and adapt to climate change.

Ironically, while there is a long way to go for solar power in Malaysia's generation mix, the country is among the top solar panel manufacturers in the world. Most of what is produced ends up being exported. International players—such as Q-Cells, which I mentioned in my speech in 2008—come to Malaysia for the attractive government policies and low cost of skilled labour. Additionally, the increasing rivalry between China, the world's biggest solar panel manufacturer, and the US and Europe has opened opportunities for players from all three powers to make Malaysia a manufacturing destination.

There is a lot of focus on large-scale solar or utility-sized solar plants, otherwise known as solar farms, where solar, photovoltaic panels are planted across large areas on the ground to generate electricity. This includes floating solar, where solar panels are placed on water bodies. The placement of floating solar facilities on water reservoirs of hydroelectricity dams presents a unique opportunity. There is a pre-existing connection with the transmission system due to the hydroelectric dam. During the dryer months, the solar panels can mitigate the reduction in electricity generated from the dam. At night, when the solar panels do not generate any electricity, the dam continues to produce electricity. In Batang Ai, Sarawak, a floating solar facility is being developed. In 2020, a 13 megawatts peak floating solar installation was commissioned in a former mining lake in Dengkil, Selangor. Floating solar helps to ease the demand for land.

But another alternative to overcome land scarcity is rooftop solar. In Australia, rooftop solar contributed more than a quarter of the renewable energy generated in 2022.[29] The electricity generated is almost double from large-scale solar. One in three Australian households have solar panels installed.

Grid, Grid, Grid

In my speeches on the energy transition, I enjoy repeating a quote by John Gummer, the former Chair of the UK's CCC and a well-regarded former secretary of state for the environment. He said that to achieve net-zero, the UK 'needs to face up to three words . . . It's grid, grid, grid.'[30]

The local grid needs to be upgraded to handle more renewable energy as well as greater electrification. Part of our net-zero objectives will only be achieved with low-carbon mobility: the adoption of electric vehicles ranging from electric lorries, busses, and cars. For electrification to truly promote low carbon mobility, the grid must be green.

But large-scale solar power requires enormous amounts of land, and it is still intermittent in nature. One would think that due to our climate and location, we would enjoy among the longest peak sun hours. In fact, for Malaysia, that period only ranges between four to 5.4 hours a day, depending on the exact location (Sabah and the north of the Peninsula enjoy the longest peak sun hours in the country). Tropical climates also face the challenge of cloud cover. The northern Peninsula, interior of Sabah, and interior of north-east Sarawak have the best sun irradiation in the country. Battery storage and pumped energy storage are options

---

[29] Clean Energy Council, 'Clean Energy Australia Report 2023', 2023, p. 7, https://assets.cleanenergycouncil.org.au/documents/Clean-Energy-Australia-Report-2023.pdf [Accessed September 21, 2023].
[30] Cooper, Charlie, 'Britain's Creaking Energy Grid Isn't Ready for Net-Zero', *Politico*, August 4, 2023, https://www.politico.eu/article/uk-energy-grid-clean-power/ [Accessed September 21, 2023].

to overcome the intermittency, as intermittently generated power can be stored to be discharged when needed.

Some of these solar farms are located further from demand, thus requiring more transmission lines that can handle higher voltage and minimal transmission loss. Traditionally, laying power transmission lines through residential areas is always controversial due to the nimbyism—the not-in-my-backyard attitude—of some communities. Direct exposure to electromagnetic waves emitted by high-voltage cables is a health risk, but it is generally established that as long as a minimum distance is maintained, the risk is minimal for the public. In Malaysia, properties next to power transmission lines are generally 10 per cent cheaper than equivalent properties further from pylons.

Going back to the energy trilemma, for many years, renewable energy only addressed one part of the trilemma—sustainability. Other forms of energy were seen as more attractive with regards to cost and uninterrupted availability. As a producer of oil and gas, diesel and natural gas initially ticked these boxes for Malaysia.

But prices of fossil fuel have shot up and have fallen substantially for solar panels. In 2010, the price of solar panel per watt was US$2.15 per watt. Eleven years later, the price plummeted to US$0.27 per watt! Without a significant new discovery, ASEAN has been predicted to become a net importer of natural gas as early as 2025.[31] Malaysia is the world's fifth biggest LNG exporter in 2022, after Qatar, Australia, the US, and Russia. But, as the demand for natural gas grows with our policy of no new coal plants, it is expected that Peninsula Malaysia will be a net importer of LNG by 2030. Thus, strengthening our renewable energy sources does not only address *sustainability* and *affordability*, but is a matter of energy *security* as well.

---

[31] ACE, 'The 7th ASEAN Energy Outlook 2020-2050', 2022, p. 79, https://aseanenergy.org/the-7th-asean-energy-outlook/ [Accessed September 4, 2023].

The Malaysian government introduced the GET scheme in 2022 to match the renewable energy generated in Peninsula Malaysia. The power supplied is bundled with renewable energy certificates. Customers are charged standard tariff rates like other customers, but instead of paying for ICPT (which is the pass-through of prices of coal and gas), they pay the GET subscription charges. Almost 2,000 TNB customers were offered and snapped up 4,500 GWh. The GET charges were increased to 6,600 GWh in January 2023. In July, I took the decision to push for a six-fold increase for GET from 3.7 sen/kWh to 21.8 sen/kWh and, again, attracted many criticisms. Some asked, shouldn't GET be cheaper, as fuel prices have increased substantially while prices of solar panels have dropped? Even if the price is pretty much the same, why shouldn't the government subsidize green electricity as it used to provide blanket subsidy for brown electricity? This allows more people to subscribe to GET.

The picture is slightly more complicated than that. As mentioned above, there are massive investments needed to upgrade the grid in preparation for the scaling up of renewables. As mentioned earlier, Vietnam had an impressive boom in solar power capacity, but then the renewables sector was beset by curtailment: solar power producers were forced to turn-off their panels even during peak hours, as the grid was unable to accept it. The UK, which, like Vietnam, has seen impressive progress with regards to renewable energy, suffers a similar problem in terms of the national grid being behind the exponential growth in green energy. Unlike traditional power generation, which comes from a smaller number of large power plants with big capacity, renewables are located across the country, supplying small amounts of power from many sites. Each new project requires connection points and larger substations. The grid must be strengthened. As of 2023, some projects must wait for fifteen years for the grid to be ready.

Then, the renewable energy certificate bundled with GET is very much sought after by companies that want to or must source renewable energy, particularly among the RE100 companies that have commitments for transitioning fully to renewable energy. Rather than burdening the average consumer, who is focused on affordable electricity with the cost of improving the grid, it is only right that customers who are willing to pay for it are charged. Otherwise, with an ongoing cost-of-living crisis, we will only invite a backlash on our energy transition journey.

ASEAN Power Grid

The previous government had banned the export of renewable energy in October 2021. This happened because the demand for renewable energy increased as Singapore wanted to achieve its own net-zero targets and attract RE100 companies. The logic to taking a protectionist approach was that as we still need to accelerate the share of renewable energy, we should not allow our green electrons to be sold to our neighbours. I believe that the memory of Malaysia losing out on the Johor River Water Agreement 1962—as part of which, Singapore was able to buy raw water from Johor for cheap—haunted many policymakers on this matter. Many officials made strong arguments to uphold the ban.

But the counterargument was straightforward: we would be able to export to Singapore, which was willing to pay a premium on renewable energy, much more than what the Malaysian market was willing to pay. Singapore had been talking about importing renewable energy from Australia, Cambodia, Indonesia, and even Sarawak. Why should Peninsular Malaysia—which, at the Johor-Singapore Causeway, is just one kilometre away from Singapore—not take part in supplying renewable energy to the city-state? We do not have to export everything, but even selling a portion of green energy to Singapore allows Malaysia to scale up renewable energy generation, invest in newer technology—such as battery storage that would be too expensive otherwise (yet necessary in the scaling up of renewable energy)—and upgrade the grid to increase renewable energy in the network. We can protect our national sovereignty and strategic interests and reach our net-zero goals while allowing for cross-border electricity trade to take place with a proper mechanism in place.

For Malaysia, in 2023, the estimated cost to achieve its net-zero target by 2050 was a total of RM180 billion, or almost RM7 billion a year. Without allowing for the export of renewable energy, the entire cost to upgrade the grid falls on customers in Malaysia. By allowing the sale of green energy to Singapore, we can use the premium to make the necessary improvements to our grid for greater supply of renewable energy. In May 2023, barely six months into my tenure, Economic Minister Rafizi Ramli and I announced that the lifting of the ban on renewable energy export. I repeated the merits of this policy every time the issue was raised in Parliament.

After all, Southeast Asia has long envisioned the creation of the ASEAN Power Grid. This will connect member states to enable mutual support and energy trading. Having interconnections across ASEAN would help ramp up renewable energy by mitigating the intermittence of renewable energy. As mentioned above, countries near the equator, such as Malaysia and Singapore and much of Indonesia, are not ideal locations for wind power. Countries further north, such as in Laos or Vietnam and the Philippines, on the other hand, can (and already are) harnessing wind energy. But wind itself is variable in nature. Similarly, the peak time for solar is about four to six hours and weather conditions also influence whether there is sufficient sunlight to generate electricity. There are also unique demands across countries due to their different economic profiles and time zones. More interconnections will address this issue, allowing countries to sell surplus energy to countries in need, enabling the scaling up of green energy.

Laos has already been supplying Malaysia and Singapore electricity generated from its hydroelectricity dams via Thailand and then through Peninsular Malaysia–Thailand interconnection.

Since 2016, Sarawak has already been supplying electricity to Kalimantan, Indonesia, and an interconnection between Kalimantan and Sabah is being studied. There is a proposal to connect Sumatera in Indonesia to the Peninsula, in what is expected to be the first cross-border submarine interconnection in the region. If ASEAN is to be part of Malaysia's future then the ASEAN Power Grid is an integral part of it.

In fact, one of the enablers of impressive renewable energy growth in Western Europe, including in the UK, is the interconnection between the countries there. A study has shown that compared to isolated grids, interconnected ones improve the stability of the grid as well as lower energy costs.[32]

Hydrogen

While airships have been around since the nineteenth century, it was from the turn of the twentieth century until the eve of the First World War that it grew rapidly as a mode of transport. Most were floated by hydrogen. German General Ferdinand von Zeppelin created rigid airships, powered by Daimler engines. From June 1910 to July 1914, as Europe was about to go to war, the world's first airline, DELAG, successfully operated Zeppelin airships that carried over 34,000 passengers in nearly 1,600 flights and travelled over 173,000 kilometres. During the First World War that ensued, Germany, Italy, and France used airships for military purposes.

In May 1937, one of these Zeppelin airships, the Hindenburg, left Frankfurt for a round trip over the Atlantic to Lakehurst, New Jersey—just outside New York City. At that time, while airplanes were around, they were still not able to carry passengers

---

[32] Jacobson, Mark Z. 'The Cost of Grid Stability With 100% Clean, Renewable Energy for All Purposes When Countries Are Isolated Versus Interconnected', *Renewable Energy*, vol. 179. 2021. pp. 1065–1075. https://www.sciencedirect.com/science/article/pii/S0960148121011204 [Accessed September 6, 2023].

over the ocean. As the airship was about to arrive in New Jersey, a storm awaited it. It had to turn and fly over Manhattan, attracting New Yorkers to rush and see the Titanic-sized airship as it passed through the city's skyscrapers. As the storm subsided, it turned back to head towards Lakehurst. As it was about to land, the airship caught fire resulting from a spark that lighted hydrogen that was leaking from the envelope. The fire spread and then the tail began to shift downwards. Within a minute, most of the ship burned down. Thirty-six people died.

This incident not only killed off the airship industry as a major form of transportation but it also created a negative perception about the use of hydrogen as a fuel. There have been three other major airship accidents—two using hydrogen and another using helium. What was special about Hindenburg was that it was the first to be caught on film and shown on newsreels in cinemas at that time. This had an impact on the public. Hindenburg during the golden age of newsreels was what September 11, 2001, was in the days of 24-hour news cycle.

In fact, the fear of hydrogen's flammable and explosive nature is irrational when we consider that petroleum is even more so. It is not only lighter than petroleum vapour but also air. This reduces the risk of it catching fire on the ground. It is not toxic for our health unlike fossil fuels. Petroleum requires only 1–3 per cent oxygen concentration for it to be explosive, whereas for hydrogen it is between 18–59 per cent.

During the oil price hikes in the 1970s, Japan tried to experiment with hydrogen again. While the Japanese named energy conservation programmes as part of the Moonlight Project (see p. 22), it branded new energy research as the Sunshine Project. Other than looking into solar and wind energy, the government explored hydrogen as an alternative source of energy. Experts in the US were also looking at the hydrogen economy to address finite fossil fuels as well as environmental pollution. But hydrogen did not scale up at that time.

With today's higher safety standards and the urgency to address climate change, hydrogen is looking attractive again. As an energy carrier, it can be used in various ways: for mobility (more on that later); to heat what are traditionally energy guzzling industries: steel, cement, and glass; as well as to produce low-carbon fuel for shipping. Hydrogen can also be a baseload fuel for power generation to balance out the intermittent nature of renewable energy and replace coal and natural gas plants.

Most of the hydrogen currently in the market is grey hydrogen, manufactured from natural gas or methane. This process releases carbon emissions into the atmosphere. Even Japan, which boasts one of the most impressive hydrogen economies in the world today, is still predominantly relying on grey hydrogen. This includes importing hydrogen made out of brown coal from Australia.

That is why the target today is to produce blue hydrogen, manufactured using natural gas but coupled with carbon capture. The ideal is definitely green hydrogen, where hydrogen is produced from electrolysing water, powered by renewable energy. The only by-product of this process is water. It is technically possible to produce green and blue hydrogen today. The hurdle to making it competitive mostly is the high cost of doing so with today's technology in most countries across the world. This is particularly true for green hydrogen. The electrolysis process is extremely expensive.

However, in China, green hydrogen is cheaper than blue hydrogen. This is because the price of electrolysers in the country is more than 70 per cent cheaper than those sold in the West.[33] Off the port of Ostend, Belgium, an offshore commercial

---

[33] Martin, Polly, 'Blue Hydrogen Cheaper Than Green H2 in All Markets Except China Amid Falling Gas Prices: BNEF,' *Hydrogen Insight*, July 13, 2023, https://www.hydrogeninsight.com/production/blue-hydrogen-cheaper-than-green-h2-in-all-markets-except-china-amid-falling-gas-prices-bnef/2-1-1486049 [Accessed October 31, 2023].

hydrogen production centre is being built, powered by wind-generated electricity. When completed in 2026, it will produce four tonnes of hydrogen a day, which will be transported back to Ostend. One of the shortfalls of intermittent renewable energy, such as solar and wind, is that the energy produced does not match the demand pattern of consumers. Projects such as this allow excess electricity to be used for the electrolysis process to create green hydrogen.

MOSTI developed the HETR, which was launched in October 2023. Petronas and TNB are doing a feasibility study to produce green hydrogen. Petronas has collaborated with a local institution, the National University of Malaysia, to produce Southeast Asia's first commercial electrolyser. Sarawak has played a crucial role on this front. By 2027, two hydrogen plants in Bintulu are expected to produce blue and green hydrogen and ammonia for export purposes. Ammonia is produced from hydrogen but is more convenient to store.

Overall, Malaysia is on the right track to a just and responsible energy transition. This is proven by the World Economic Forum ranking the country first in Southeast Asia in its 2023 Energy Transition Index. We need to leverage our legacy of successfully providing electricity over the years. But, as mentioned above, challenges remain. We need to fill in the gaps by electrifying Sabah and increasing investment for overall grid readiness. The world is moving fast, and we cannot afford to be left behind. The momentum for structural reform for the electricity sector—what is being called 'futureproofing'—must continue to deal with the changing realities of more renewable energy and to make the ASEAN Power Grid a reality. These reforms are for Malaysia to continue to trade and attract investors. Ultimately, however, this is key to achieving our net-zero goals and for the survival of this planet

Transforming the Water Sector

'The next war in the Middle East will be fought
over water, not politics,'

—Boutros Boutros-Ghali,
UN Secretary General 1992–1996

Water constitutes 60 per cent of the human body, and over 70 per cent of our planet. It is central to our survival. While water is plentiful in one sense, it is also scarce. Only 1 per cent of the water available today is non-saline, fresh water while another 2 per cent is frozen as glaciers and ice sheets in the polar areas. Managing water becomes even more difficult as we face the triple planetary challenges of climate change, environmental pollution, as well as biodiversity loss.

I have fond memories of returning to my grandmother's house in Kota Bharu, not far from the banks of the mighty Kelantan river, which is one of the longest in Peninsular Malaysia. My grandmother's home was a traditional, wooden Malay house, built on stilts and bedecked with Senggora roof tiles that resemble brown fish scales. While we were growing up, my cousins and I would play under the house in the space between the floor and the damp ground—until we became too tall. We looked forward to going back during the monsoon—a migration Malaysians call *balik kampung*. In the days before AirAsia and Airbnb, only the rich would enjoy holidays abroad. For the middle classes, balik kampung was often all they had.

Image 8: The Prime Minister of the Netherlands, Mark Rutte, taking a selfie after we finished the 6th Malaysia–Netherlands Water Dialogue in Kuala Lumpur. The fact that he was tall helped in getting a nice selfie!

This often coincided with the year-end school holidays. Floods were regular events, and we would measure it based on the highest step the floodwaters would reach every year. In the space under the house, where we played, a simple boat belonging to my late grandfather was stored. When my mother was growing up, they would take the boat out during the floods and go around the submerged town, snacking on boiled yam and corn. Of course, this was a paradox—an event that brought massive material damage, even loss of life, was also a source of simple joy for children. Diseases, such as cholera and leptospirosis, also spread easily due to poor hygiene during floods.

Elevated houses have been a typical feature of homes in Southeast Asia—an early form of climate adaptation to deal with floods, wild animals, as well as providing better ventilation. Wooden houses on stilts could be physically transported to another location. After all, Malaysia is a tropical country that experiences heavy rainfall throughout the year, making floods a frequent occurrence, even in Kelantan. Modern homes are now built on the ground as traditional wisdom is discarded. I do not have an answer for why this happened: whether it is our hubris in conquering nature or a belief that Western architecture is synonymous with modernization; it is anyone's guess. However, Kelantan, besides the floods, also faces problems of water quality and supply. For one thing, the state is more dependent on groundwater than any other state in the country. The water extracted contains high levels of iron and manganese, and occasionally even arsenic.

I am often asked—how and why does this happen? When it comes to water, we transition from abundance to scarcity in rapid cycles. We suffer from floods in some months and droughts in others. This is the triple whammy of water that is often repeated—water being too much, too little, or too dirty.

One day, the daughter of a rubber tapper from a kampung less than 100 kilometres away from Kuala Lumpur told me an interesting story. She recalled how she could drink water directly from the well when she was growing up. Who needs expensive bottled mineral water from the Alps when one can source it from their doorstep? After finishing her university studies and starting work in the civil service, she brought her father to the city. Her father was surprised at the fact that everyone was using so much bottled water for drinking. Why was this necessary, he asked. He told her that if nothing was done, would we also have to pay for oxygen canisters to breathe in the future? Let's also not forget about the plastic problem, which I will touch on in more detail later.

I visited the Orang Asli community in Gua Musang, near the highlands of Kelantan, in November 2023. A group of Kelantanese government officers and community leaders joined me on the trip. The Orang Asli live on the banks of the Nenggiri river. I was aghast by what I saw. We got on bamboo rafts and slowly went downriver to reach them. We stopped at Kampung Podek to plant trees and provide assistance. They had water pipes that reached their village, but they could not access treated water. I saw mothers and their children bathing, washing, and playing in the river. What made it worse was that the vast jungle that used to line the banks of the river had been cleared for oil palm and durian plantations. The river that used to be clean is now murky. This Nenggiri river flows into the Galas river and then into the *teh tarik* (local 'pulled' milk tea) coloured Kelantan river—where my grandmother's house stands, in Kota Bharu.

There are certainly things we can be proud of when it comes to water in Malaysia. We have succeeded in providing high overall access to potable water supply—over 97 per cent in 2022.[34] Our sewerage services have reached over 85 per cent in the main cities. Malaysia has embarked on various innovative ways to deal with the flooding issue, which I will elaborate below in the section on flood mitigation. Water supply is now more focused on performance, while privatization, which failed miserably, has been rolled back.

But many challenges remain. There are still pockets of communities that are not able to access piped water and that's where we need to focus. With climate change, worsening floods and coastal erosion will become bigger problems for the country.

Water Management in Malaysia

I visited a traditional religious school, the Madrasah Ihya Ulumuddin in Pasir Puteh, in Kelantan in 2023. I discovered that the institution, which teaches classical religious sciences, still uses texts written by my great, great grandfather, Tuan Tabal (1840–

---

[34] 'Water Malaysia 2023 Specialised Conference and Exhibition', *Malaysiakini*, June 15, 2023, https://www.malaysiakini.com/announcement/668744 [Accessed January 30, 2024].

1894), which he had composed for Sufism or *tasauf*. I was taken to the principal's house, and I was informed about the severe water supply issues there. Like many in Kelantan, they were relying on groundwater, with over twenty shallow wells being built. However, the water was brown and heavy.

Image 9: I made a visit in June 2023 to a traditional Islamic school, Madrasah Ihya Ulumuddin in Pasir Puteh, Kelantan. Like many residents of Kelantan, they complained about having to rely on poor-quality groundwater due to water supply problems. Within twenty-three days, SPAN delivered new pipes, tanks, and pumps to provide treated water.

Wearing white garb is a key marker of identity for the teachers and students there, but the principal lamented: 'We even have to send our turbans and robes to be washed in the laundry outside to keep them white.'

Initially, we thought a solution would be to build deeper wells to access clearer and cleaner water in the madrasah. But after getting feedback from SPAN, we decided that merely investing in a more powerful pumping system to tap into Air Kelantan's water supply would be a better alternative. SPAN decided to do a CSR programme by building the necessary infrastructure for the madrasah.

As noted, the water supply issues in Kelantan have been longstanding, although they seem to have gotten worse in recent years. In 2023, Kelantan was the worst state in Peninsular Malaysia in terms of compliance with the indicators set by SPAN. It is not limited to the rural areas (as I described via the problems faced by the Orang Asli in Gua Musang or the madrasah) but can be seen across the entire state. As recently as 2021, my predecessor in charge of water supply, Tuan Ibrahim Tuan Man, replied that only 71.7 per cent of Kelantan residents had access to clean water (remember, at the national level, the rate is 97 per cent). A substantial number of water treatment plants were not running at capacity. This, combined with the high NRW rates, meant that the Kelantanese have no choice but to use groundwater.

Groundwater is a widely used source of water globally. It is the source of half of the world's drinking water (similarly, half of the drinking water of the US) and 40 per cent of the water used for irrigation. The problem is that while it may be suitable for drinking in some areas, it is not so in others.

I remember when I was studying in London, how different tap water tasted from the water we were used to in Malaysia. This is because more than half of Britons (including Londoners) rely on hard water, high in mineral content, which comes from groundwater, drawn from limestone aquifers in the ground. The water in the UK has high levels of magnesium and calcium. At the levels consumed in the UK, it is safe to drink. But aside from the strong taste, it also created limescale problems for our kettles!

In Kelantan, however, the level of iron and manganese is very high in groundwater. In fact, at times, there are also high levels of the toxic element arsenic. For the water to be drinkable, a lot of money needs to be spent on its treatment.

As a result, many Kelantanese resort to bottled water and drinking water vending machines. Not only is this bad for the environment, but it is also expensive, particularly for those who live far from urban centres. It gets worse during school holidays and Hari Raya, as many of the Kelantanese who are working and studying outside the state come back to the state, excessively burdening the limited water supply. Real-estate developers are being forced to provide mini water-treatment facilities on their premises, which adds to the costs. This deters many investors from coming into the state, contributing to the vicious cycle of underdevelopment.

Traditionally, water was managed on a fragmented basis, as it was always part of the State List under the Federal Constitution.[35] The Federal Government, however, allocates funding for water projects. In the early 1990s, just like in the electricity sector, several states privatized the water sector. The move started with water treatment plants because this was seen as more profitable by private investors unlike distribution. But the EPU then deemed that the only efficient way for privatization of the water sector was for all three parts—catchment, treatment, and distribution—to be privatized.[36] IWK was created in 1993 to take over sewerage services from local governments.

In the UK, or rather England and Wales (because Scotland and Northern Ireland authorities continued to hold on to water and sewerage services), privatization of regional water authorities was completed in 1989. Perhaps aptly, North West Water, one of the privatized regional water authorities, owned one of the companies that was part of the IWK consortium. Similarly,

---

[35] The Federal Constitution spells out that Parliament has the authority to make laws under the Federal List, the state legislatures make laws under the State List while the two legislatures have shared authority over the Concurrent List.

[36] JICA, 'A Study of Privatisation in Malaysia', 1999, p. 7–17, https://openjicareport.jica.go.jp/pdf/11634078.pdf [Accessed August 9, 2023].

another privatized English regional water authority, Thames Water, used to own 70 per cent of Air Kelantan. On the other hand, YTL, a Malaysian infrastructure conglomerate, has owned Wessex Water in the UK since 2002.

This interesting relationship can probably be traced back to the fact that being part of the Commonwealth, Malaysia models a lot of its policies and laws on the UK. In fact, Malaysia's privatization was modelled on UK's own experience initiated by Margaret Thatcher in the 1980s. In Malaysia's reform of the water sector later in the 2000s, which I will elaborate further below, the UK was again used as one of the points of reference for the new legislation.

Soon, however, the flaws of privatization for water and sewerage services became apparent. Seven years after IWK's creation, the government decided to bail it out. IWK changed hands four times in those seven years while the government had pumped RM1 billion into it during that period. It had a RM700 million debt with the government, and consumers complained that the municipalities that used to run sewerage services were more responsive than the supposedly efficient private concessionaire. The nationalization (in effect, re-nationalization) of IWK in 2000 included a RM200 million compensation to IWK's holding company.[37] It was another expensive error at the cost of the taxpayers.

This was then extended to state water operators. The massive capital expenditure required for the water supply meant that the operators had to take big loans. Low repayments led to government loans becoming limited, and some operators resorted to borrowing from the market. They wanted to pass through the costs, via tariff increases, to the public. The state governments

---

[37] Chong Yen Long, 'IWK Bailout Shows Perils of Privatisation', *Malaysiakini*, April 19, 2001, https://www.malaysiakini.com/news/2133 [Accessed July 31, 2023].

were not keen on this for obvious reasons. The business model simply did not work. It turns out that not everything can be passed to the market, and if the market fails, the government must still step in.

In 2005, the Federal Constitution was amended so that while the control of water sources remained with state governments, the Federal Government would thus forth regulate water supply and services. It was helpful that the BN government had two-thirds majority in Parliament and controlled twelve out of thirteen states in the federation then. Regardless, as I learned from the civil servants in NRECC who dealt with it at that time, getting the changes through was still challenging. Today, in an era of greater political competition nationally, and with many state governments run by the national opposition, such an amendment would be a tall order.

The tide turned against privatization globally as its limits became clear. Political and environmental opposition meant decisions to privatize water operators were subsequently reversed in cities such as Berlin, Johannesburg, Paris, Accra, and Buenos Aires—they have all 're-municipalized' their water systems in the past decade. This was clearly an area where the market failed.

In England, which is one of the few countries to adopt wholesale water privatization, water operators were deemed to have failed miserably. Critics argue that it led to increasing bills and underinvestment. In 2022, 301,091 spills by English water operators were recorded—more than 800 in a single day! Water pollution became one of the major issues in English local council elections in 2023.

In July the same year, the UK's largest water operator, Thames Water (a former major shareholder in Air Kelantan) was close to collapse after years of massive borrowings—much of which were funnelled towards unsustainable profits for its owners and lavish pay and bonuses for their bosses instead of reinvesting in much-needed infrastructure. As a result, it was fined millions of pounds for environmental offences. One of Thames Water's treatment plant discharged two billion litres or 400 Olympic-sized-swimming-pools-full of sewage in a period of two days in 2020.[38] As the company was looking at becoming insolvent, the Conservative government—the so-called ideological heir to Margaret Thatcher—was considering taking it back to public ownership, at least temporarily.

Privatization was worse in Malaysia, as it became an excuse to award comfortable contracts to politically connected individuals, which was dubbed as 'crony capitalism'. Mahathir's fetish for privatization combined a zeal to emulate Thatcher and Ronald Reagan, with authoritarian tendencies as well as the justification of a need to address racial imbalances in the economy. As in so many tragedies with regards to privatization, the profits were privatized, but the losses were socialized.

In Malaysia, low water tariffs contribute to inefficient consumption. Without having to pay the real cost of water, the public used more water than needed. In 2022, our water consumption was 261 litres per capita per day. The UN recommendation was almost 100 litres less, at 165 litres per capita per day. Every Malaysian is basically consuming over 400 more cups than necessary on a daily basis.

One of the items mentioned in the 2008 Selangor BA manifesto was to provide free water to households in Selangor. This was the first election I contested. To be honest, I myself did not imagine I would be able to win my seat, let alone that the opposition (BA and DAP) would form the state government. If we did, we probably would not have agreed to this pledge. This is a perennial lesson for politicians—manifestos must be properly

---

[38] 'River Thames: More Than 2bn Litres of Raw Sewage Discharged Over Two Days', BBC, January 18, 2022, https://www.bbc.com/news/uk-england-london-60046320 [Accessed July 11, 2023].

thought through. There has to be a sense of responsibility, it must not merely be clickbait that will draw publicity.

We won, and as I was also the political secretary to the Menteri Besar Abdul Khalid Ibrahim, I attended meetings with the state civil servants who pored through our manifestos and discussed priority policies. In June, three months after the election victory, Khalid announced the free water policy of 20 cubic metres to households in Selangor. It was a popular (and populist) policy.

I remember my father, who is a retired senior government servant, being pleased about the savings (RM11.40) to his water bill. Indeed, the policy was popular for the poorer households, as a saving of RM11.40 meant a great deal for low-income earners. However, in reality, the same amount meant little to middle and high-income households like mine. For the state government on the other hand, this added up to about RM131 million in spending per year for the first full year of 2009 and reached almost RM200 million in 2018. From 2008 to 2018, RM1.8 billion was spent on providing free water. It did not encourage responsible use of water and, as it applied equally to all households, was a regressive policy. In 2019, the Menteri Besar Amirudin Shari changed the policy whereby only registered households earning less than RM4,000 a month would be the beneficiaries.

The problem is not unique to Selangor. As of 2023, Pahang has not increased their water tariff since 1983, a year after I was born. This was Bob Paisley's final season as Liverpool manager, right in the middle of a three-year consecutive League victory for the Reds. The hit song for that year was The Police's 'Every Breath You Take'. Meanwhile Perlis has not raised tariffs for twenty-seven years. In 2022, the average water tariff charged in the Peninsula and Labuan was RM1.38 per cubic metres whereas the capital and operational costs were RM1.68 per cubic metre. This resulted in leaking pipes and a lack of maintenance—aggravating the NRW issue. Again, as with electricity tariffs, the idea of maintaining a low-cost economy—which includes paying our workers low wages—was the fundamental flaw in the economic structure that allowed us to suppress our water tariffs.

During my year as the minister in charge of Malaysia's water supply, I focused on getting the tariff-setting mechanism for the water sector in place for domestic customers. We managed to get the state governments across the party lines to agree on the need for this. The previous government had allowed for an increase for non-domestic customers in 2022, but this had negligible effect on the bottom line of water operators in less developed states where the number of commercial and industrial customers is small. I managed to bring the tariff-setting mechanism to Cabinet at the end of 2023—my last act as the minister responsible for water and energy (i.e., in the transition from NRECC to NRES).

Image 10: Inspecting newly installed solar panels at a sewage treatment plant owned by Indah Water Konsortium in Kuala Lumpur.

The previous government had also decided to increase the tariff for sewerage services by IWK starting 2023. The last tariff revision was done in 1994—when I was still in primary school! Despite public relations efforts by the Ministry of Environment and Water to alert the public about this hike, we received a lot of complaints when the public saw the increase in their bills after I took over. The cost for connected sewerage treatment was about RM17 per month, but houses connected to the sewage services only experienced a mild hike from RM8 to RM10 per month—a mere RM2.

Charging water and waste water tariffs closer to the market rate makes rainwater harvesting and water recycling more economical. Rainwater harvesting collects and stores rain, instead of letting it run off. This has enormous potential in Malaysia due to the heavy rainfall we receive. Similarly, water recycling, as well as greywater and stormwater recovery, has enormous potential to reduce the burden on potable water. Wealthy countries where access to water is limited—such as those in the Middle East, like UAE and Qatar, as well as our neighbour Singapore (where much of the water supply comes from Malaysia and has historically been a thorny political issue)—have long explored all these measures.

IWK began water recycling measures since 2015. Treated waste water, called bioeffluent, is used by industries and for landscaping. The company has three water reclamation plants in Klang Valley, Malacca, and Penang.

There are other innovative methods we could consider. In Malaysia, the Water Waqf programme has been established for the benefit of those who have yet to get water supply. *Waqfs* are Islamic endowments that are a popular form of charitable instrument in Muslim communities. Muslims believe that upon death, all their good deeds no longer benefit them except for charitable donations, beneficial knowledge that continues to be taught and practised, and the prayers of their children. Waqfs have spawned institutions of learning, hospitals, soup kitchens, water wells, and drinking fountains.

The Malaysian Water Waqf programme involves small-scale, off-grid projects such as tube wells, water pumps, and water tanks. Both one-off grants as well as maintenance allocations are provided under the initiative. These allow the public and corporations to donate to the programme, benefitting tens of thousands of Malaysians. While this may seem a small initiative compared to what is required to address the shortfall of water supply in the country, it can be an alternative and complimentary effort towards providing the universal supply of clean water.

Non-Revenue Water

From a state-by-state perspective, some suffer from higher NRW rates than others. NRW is the percentage of treated water that is lost before reaching consumers—through leaks, thefts, or inaccurate water meters. The national average was 37.2 per cent in 2022. In other words, more than a third of water that is treated does not reach consumers. This results in a loss of RM2 billion a year![39] Reducing NRW is possible but, traditionally, challenging due to a few reasons.

In Malaysia, there are two approaches to overcome the issue of NRW. The first approach is for states with NRW above 40 per cent. Here, the government directly provides grants for them to conduct NRW reduction projects. The second approach is for states with NRW below 40 per cent, where SPAN provides a matching grant. For every ringgit spent by the water operator, SPAN pays them a matching amount. Prior to my tenure, this was done when the operators reached a certain agreed level, say 2 per cent per year. But when they fell short—if they were only able to reach 1.5 per cent—they would not receive a single ringgit. This deterred many of the water operators who were not confident of reaching the target and were worried about

---

[39] 'Dewan Rakyat Debate', October 25, 2023, 60, p. 15.

coming up with capital expenditure. Many water operators were not bankable due to their financial situation. NRECC improved this by allowing the matching grants to be provided pro-rated to what has been achieved. So, as with the example above, the water operator would still receive 75 per cent of the total matching grant for the capital expenditure.

Shortly after taking office, I spoke to a former water minister from a neighbouring country who told me something that I felt compelled to repeat in my speeches: politicians like opening shiny new treatment plants where they can showcase a tangible and expensive megaproject to the public. In contrast, changing pipes or cutting illegal connections does not involve ribbon-cutting photo-ops that look good on Instagram. This is similar to the energy sector where opening new solar farms is politically sexier than improving energy efficiency, though both are needed.

But, according to the World Bank, politicians alone are not to be blamed for shying away from solving the NRW issue:

> Physical loss reduction is an ongoing, meticulous activity with few supporters among the following [. . .] Engineers: it is more 'fun' to design treatment plants than to fix pipes buried under the road. Technicians and field staff: detection is done primarily at night, and pipe repairs often require working in hazardous traffic conditions. Managers: it needs time, constant dedication, staff, and up-front funding.[40]

Yet, simply channelling more money to address NRW alone is insufficient. We must make sure that water operators are managed

---

[40] Kingdom, Bill, Roland Liemberger, Philippe Marin, 'The Challenge of Reducing Non-Revenue Water in Developing Countries—How the Private Sector Can Help: A Look at Performance-Based Service Contracting', *Water Supply and Sanitation Sector Board Discussion Paper Series*, no. 8. World Bank, Washington, DC., 2006, p. 17, https://openknowledge.worldbank.org/bitstreams/93c4ed89-423f-5053-a553-9e181e595b78/download [Accessed June 19, 2023].

well and efficiently. This ensures that the operators properly spend the money allocated to them. Otherwise, we are just throwing good money at something bad.

River Clean-up

Malaysia's greatest conurbation is the Klang Valley, home to nine million people. It takes its name from the Klang river, which begins in the Titiwangsa mountain range, near the iconic Klang Gates Quartz Ridge in Selangor, crosses the Federal Territory boundary into Kuala Lumpur and then back to Selangor. Towns grew there during the tin mining boom as the Klang river became not only a source of water but also transportation. Rapid development has led to massive pollution due to industrial effluents and poor sewage management. Heavy siltation, narrowing riverbanks and an increase in impervious surfaces cause flash floods. Because of this, the Klang river has been dubbed Malaysia's dirtiest and was once listed as one of the fifty dirtiest rivers in the world. Due to this, the river's water cannot be used for water supply, and the Klang Valley has to rely on the cleaner Selangor river further up north for this.

There have been various projects to clean up the river, including the Federal Government's River of Life and the state government's Selangor Maritime Gateway. As minister, I had the chance to see projects, big and small, that have led to an improvement in the water quality. I visited the Interceptor 005 floating barrier, sponsored by the famous British band Coldplay, which was deployed along with the Interceptor 002 by the Dutch non-profit Ocean Cleanup. The two interceptors were deployed in the Klang river due to the level of pollution there. Klang was among the earliest sites for the project. These solar -powered barges remove waste floating along the river. Connected to the Internet, the interceptors send real-time data about their performance. New log booms were also installed across the river to collect floating debris. The ultimate objective is not only to clean the river but, in doing so, to also prevent the plastic waste from reaching the sea.

Image 11: I visited the Interceptors stationed in the Klang river to prevent debris (especially plastic) from going into the sea. This device by the Ocean Cleanup was funded by the popular rock band Coldplay.
Photo credit: Mohd Johari Ibrahim / Bernama

On the same day, I dropped by Mangrove Point, a 70-acre area combining a permanent forest reserve and state land, now designed to be an ecotourism attraction. As it was low tide,

I could see first-hand the litter trapped in the trees. In 2023, the Federal Government signed a memorandum of cooperation with Ocean Cleanup to expand this project to the whole country due to requests from various other state governments.

Today, Malaysian rivers face various sources of pollution. As mentioned, this includes sewage, waste water, effluents from illegal factories, grease from workshops, rearing of seafood, faeces, and carcasses from animal farms, as well fertilizers and pesticides from oil palm plantations, among others.

But local communities have been able to find solutions. A river trail was also built along one of Klang river's tributaries, the Penchala river in Section 14, Petaling Jaya. This was my backyard, where my parents live and where I grew up. I passed by the stream where I used to cycle from my house to the Jaya Supermarket as a teenager. I always thought that the stream there was a monsoon drain and never thought much about it. After all, it was inaccessible to the public. However, the construction of the proper trail, which allowed the public to jog and cycle there, created a sense of ownership. In fact, prior to my visit, a member of the public who used the trail immediately alerted the authorities when they noticed whitish, bubbly water, suspecting that it was pollution.

Today, the Klang river water quality has, at times, gone up to Class III or II (just a class below being safe for contact for recreational use) from its previous position, Class V (the water is not safe for any use). During the rainy season, the water quality reaches the best level of Class I. Crocodiles and otters are spotted in the river. Over 86,000 metric tonnes of rubbish have been collected since 2016—equivalent to the weight of 470 Boeing 747s.

In Kelantan, I visited the Pulau Tengah floating market on the Pengkalan Datu river. The public can buy food, including fresh seafood, from boats floating by the river. Nearly 5,000 visitors patronize the markets every month. Traditional kite and top-making activities, popular in the state, take place along the market. You can also ride colourful traditional boats adorned with dragon motifs, the *perahu kolek* on the river. A transformative project has been the one-kilometre river trail made up of wooden planks alongside the Nibong palm trees that grow alongside the river. If only its water supply could be better managed!

One of the most well-known examples of river clean up internationally is the Han river that passes through Seoul, South Korea. As South Korea rapidly developed and industrialized in the aftermath of the Korean War, the river was severely contaminated by discharges from textile factories, sewage plants, and settlements built along the river. Fish lay dying on the sides of the river. One of the streams that feed into the Han, the Cheonggyecheon, was covered up with concrete while an elevated highway was built on top of it. In the years leading up to the 1988 Olympics, the government began a massive clean-up exercise. This included building new sewage treatment plants, dredging, and lastly, beautifying the riverbanks. Bicycle lanes, pedestrian pathways, and beautiful parks dot the banks of the Han.

In 2003, the Mayor of Seoul Lee Myung-bak decided to demolish the highway on top of Cheonggyecheon. The stream was restored. Not only did various species of wildlife return to the stream, but it also managed to lower the temperature of the area by 3.6 degrees Celsius compared to surrounding neighbourhoods in the city. It restored pedestrian paths while reducing traffic going into the city centre.

Korea's experience is illustrative of how to sustain a restoration project in the long run. The continuous cycle of improvement and maintenance is something that, unfortunately, Malaysia has repeatedly failed at in the past.

Flood Mitigation

In 2014, massive floods hit Peninsular Malaysia and Sabah. Kelantan was badly hit. People died and looting ensued as residents searched for ways to survive. While water was plentiful, drinking water was not. Mineral and drinking water were sold at a premium. A mother reportedly had to mix formula with rainwater for her six-month old baby out of utter desperation.[41] I was just a few months into my time as the Leader of the KEADILAN Youth Wing, and I worked to raise funds online and went down there with a busload of volunteers. I even dropped by my grandmother's house, which was totally deluged with floodwater just a few days before that. This was probably the highest level it had reached in the history of the house, which was more than seventy years old. At the peak of the flood, some of my relatives took a boat to go to the other side of Kota Bharu to rescue family members.

Just seven years later, another major flood hit. Rain fell heavily across Peninsular Malaysia. It continued for over twenty-four hours and broke several meteorological records. In that short period of time, the rain that fell was equivalent to the monthly average. Normally, this level of rain is only witnessed along the east coast. In Selangor, the daily rainfall recorded was more than double the highest recorded level. Experts described the chances of it happening in any given year as 1 per cent, otherwise known as a 100-year flood. The Klang Gates dam, Malaysia's oldest water supply dam located at the source of the Klang river, as well as the Batu dam had to open their gates and release a quarter of their contents. As a result, the banks of four main rivers in Kuala Lumpur were breached. A few key water treatment plants supplying the Klang Valley had to stop or reduce operations.

In Taman Melawis, Klang, water rose fast in the old housing estate. The drains from the 1960s could not cope with more intense and frequent rainfall. The residents were not strangers

---

[41] Chi, Melissa, 'In Kelantan, Desperate Flood Victims Loot Homes for Food and Fresh Water', *Malay Mail*, December 27, 2014, https://www.malaymail.com/news/malaysia/2014/12/27/in-kelantan-desperate-flood-victims-loot-homes-for-food-and-fresh-water/808959 [Accessed January 2, 2024].

to floods before and had all sorts of small pieces of engineering ready to deal with them. But this time, those measures were not sufficient. The psychological impact was so bad that some of the residents heard phantom rain—imagined sounds of rainfall induced by the anxiety and fear of flooding.[42] The state government started a flood mitigation project there, and in 2023, I decided to visit to see the progress of the project, get a briefing from the engineers responsible, and also to listen to the grievances of the local residents.

Image 12: Visiting a flood mitigation project in Rantau Panjang, Kelantan and listening to the complaints of the local community. We made instant adjustments to address some of the interim problems faced by the local community that bore fruit within six months.

Floods reduce the prices of homes. Ordinary families often get stuck with their properties, which tend to be the only ones they

---

[42] Razak, Aidila and Arulldass Sinnappan, 'When the Water Rises: A Malaysian Climate Change Story', *Malaysiakini*, August 29, 2022, https://newslab.malaysiakini.com/climate-change/en/ [Accessed October 23, 2023].

have. Who would want to buy such properties with the flood risks? How can they move out to better properties? This creates a vicious cycle. That is why calamities impact the poor and lower middle classes more severely than the rest of society.

Here's another case study. Taman Sri Muda is a low-lying housing area 25 kilometres by road from Kuala Lumpur. It lies at the meeting point of the Klang and Damansara rivers. With a catchment area of 1,100 square kilometres from Ulu Klang, Ampang, Batu, and Damansara, it is naturally suited to be a flood plain. It was originally opened as a rubber and oil palm plantation, known as the Merton Estate. To make matters worse, on 19 December 2021, Klang was also facing tidal floods due to an exceptionally high tide caused by the appearance of the new moon. Residents of two-storeyed houses could run to the top floors, but those residing in single-storeyed houses were not as lucky. The area was submerged under as much as 4 metres of water and sixteen people died. The state government had been trying to deal with flooding since as far back as 1995. To protect the residents from tidal floods, tidal gates were built to protect against backflow, pushing it upstream. Pumps were supposed to transfer the water out in these situations. These mitigation measures were largely adequate at that time. But as the area has become fully developed and receives a higher intensity of rain, the old infrastructure is no longer sufficient. In 2021, the high level of water meant that the pumps' power supply was cut. As mentioned earlier, this may also be another example of our desire to build but not maintain.

Historically, the most flood-prone areas in Kuala Lumpur are at two confluences: the Ampang and Klang rivers upstream, and the Gombak and Klang rivers. The latter is where Kuala Lumpur was first established and the origin of its name—the muddy confluence. It is marked by Masjid Jamek, a mosque built by Indian Muslim traders and conveniently located near the centre of administration for British Malaya, thus allowing Muslim civil servants to pray there. It used to be the city's largest mosque until the National Mosque was completed after Independence. At the turn of the twenty-first century, the government sought ways to ensure that continuous rain even for three to six hours in the city would not lead to it being flooded.

What came about was an innovative infrastructural adaptation—the Stormwater Management and Road Tunnel (or SMART Tunnel) that was completed in 2007. It combines a tolled-highway to ease the notorious Kuala Lumpur traffic (and pay for the project) with a stormwater tunnel that takes water from a holding pond near the Ampang and Klang rivers' confluence. The tunnel goes under the city, consisting of three levels. Normally, the top two levels carry traffic. The bottom level brings the water out from a storage reservoir before releasing it into the river. When there is heavy rain, the top two levels are closed to traffic and carry stormwater as well. The stormwater tunnel, nearly 10 kilometres long, is the second longest such infrastructure in Asia. In February 2022, it was estimated that the SMART Tunnel has helped avoid forty floods (almost half of the floods that would have happened in the same period) and RM1.4 billion in public damages.[43]

The government has also employed other innovative measures to deal with water supply and flood mitigation together—the dual function off-river storage method like the one being planned in Sungai Rasau, Selangor. Here, water that is stored as part of flood mitigation will also be used to be processed as treated water. The off-river reservoirs mean that should there be a pollution incident in the Selangor river, water can still be processed from the reservoirs. This will help solve the conundrum of a country with heavy rainfall and floods also suffering from water shortages.

---

[43] M. Mageswari, 'SMART Tunnel Mitigates 45% of KL's Floods', *The Star*, February 12, 2022, https://www.thestar.com.my/news/nation/2022/02/12/smart-tunnel-mitigates-45-of-kl-floods [Accessed March 7, 2024].

Image 13: Visiting a flood at Bukit Tinggi, Pahang with local representatives

The challenge is keeping up with climate change. New Orleans, like Amsterdam, lies pretty much below sea level, as it was built over swamp and marshland, particularly through dykes and floodwalls. The Mississippi river cuts through the city at a higher level than the surrounding neighbourhoods. Tapping into the groundwater as well as naturally rotting marshes contributes to natural settlement of the ground. This means that the city is sinking. In 2005, New Orleans was hit by Hurricane Katrina. This destroyed the dykes that protected the city and meant that four-fifths of the city was underwater.

Jakarta, the capital of Indonesia is also sinking. The shortage of clean water supply means that residents depend on groundwater, the overexploitation of which has led to the city sinking. Today, 40 per cent of North Jakarta is estimated to be below sea level. A network of canals has been built to mitigate the problem of flooding in the city. Even then, at the rate things are going—the city sinking while the sea levels rises due to climate change—it is estimated that by 2050, only 5 per cent of the same area will remain *above* sea level! In 2019, President Jokowi announced that the country will move its capital to Borneo—the Nusantara site that is still being constructed.

In the northwest of England, two flood episodes happening in a span of six years (2009 and 2015) were calculated to be the worst in over half a millennium. While other factors cannot be ruled out, it seems difficult to say that climate change does not have anything to do with the more uncertain weather patterns seen today. The 2015 flood took place during one of the warmest and wettest winters experienced by the country, and due to the storms Desmond, followed by Eva and Frank hitting England.

Research shows that a flood of a magnitude expected to occur once every 500 years prior to the Industrial Revolution in New York City is occurring once every twenty-five years today, both due to increasing sea levels and intensity of storms.[44] Our ingenuity in engineering has allowed us to flourish before this, but it will need to be harnessed fully for humanity to deal with the challenge of increased flood risks due to climate change.

There is a limit to how much we should rely on grey infrastructure alone. We need to incorporate nature-based or blue and green solutions as well for a more sustainable and holistic approach. Restoring rivers to their original meandering paths will slow the flow of water, decreasing the risk for floods. At the same time, this will help restore aquatic life in rivers. Floodplains can be restored so that they naturally reduce the risks of floods, and when floods do happen, adapt to them in a minimally intrusive manner. This will also allow groundwater to be recharged. Building green embankments, with grass and mangroves protecting the slope as

---

[44] Reed, Andra J., et al., 'Increased Threat of Tropical Cyclones and Coastal Flooding to New York City During the Anthropogenic Era', *Proceedings of the National Academy of Sciences USA*, vol. 112, 41. September 28, 2015, pp. 12610–12615, https://doi.org/10.1073/pnas.1513127112 [Accessed December 31, 2023].

well as natural reservoirs and restored wetlands lined with plants will not only beautify the grey infrastructure but also increase water retention. Stormwater tree trenches and bioswales can be built to handle water runoff. Mangroves can be combined with dykes to protect against tidal floods. While grey infrastructure needs to be managed by the public or private sector, green adaptation measures need a strong involvement from the local community. Another modern adaptation measure is an early warning system, in Malaysia we have PRABN.

The authorities in Denmark undertook a massive restoration project for the Skjern river starting in 1987. Among other measures, the authorities have lowered the riverbanks back to their original height and reintroduced wetlands. This has allowed seasonal flooding. The river's original curving path has been restored, which slows down the flow of water and prevents serious flooding downriver. The biodiversity of the area has also increased remarkably.

Water and Climate Change

The UN General Assembly is housed in the international organization's headquarters, situated in Manhattan overlooking the East river in New York. The building was opened in 1952, when the Cold War between Second World War allies of the US and the Soviet Union was brewing, and Malaya, Sarawak, and North Borneo were still administered by the British.

In March 2023, just over three months after being appointed as NRECC minister, I represented Prime Minister Anwar Ibrahim at the UN Water Conference held there. It was only when I reached the iconic hall that I was told I would be giving our country statement from the rostrum in front of the UN seal against the gold background and green marble podium and not from where the Malaysian delegation was sitting. I had never imagined delivering a speech there. I felt a tinge of nervousness. I don't usually do well in unfamiliar territory. And this was a bit more than unfamiliar, to say the least. I re-examined my text and made new edits until it was almost my turn, much to the consternation of the civil servants and my aide who had to run back and forth with my text.

Image 14: Delivering Malaysia's country statement at the UN Water Conference 2023 at UN General Assembly Hall, New York, US. I emphasised that the climate crisis is a water crisis.

Finally, I was called up to the rostrum:

> Mr. President . . . Malaysia recognises the importance of water as a critical resource for sustainable development, economic growth, and poverty reduction . . .
>
> We need to acknowledge that the climate crisis is a water crisis. The world is facing water scarcity, inadequate provision for sanitation, as well as more intense and frequent disasters and extreme weather events. The world saw devastating disasters in 2022, the costliest being Hurricane Ian in the US and Cuba

with more than $US100 billion of damages while the Pakistan floods killed 1,739 and displaced seven million people.

The IPCC in their Sixth Assessment Report has highlighted that the most vulnerable people are often disproportionately affected and we are pushed beyond our adaptation limits. This poses significant challenges for the developing Global South. Climate change aggravates and widens the existing development gap. Water is central to addressing this issue.

We need a framework for resilience and it has to start with water, where climate, environment and development merge. The world needs to shift how development is approached in the age of crises where climate and disaster risks, and economic growth have to be planned together . . .

Every development plan must consider the water element and its impacts. We cannot afford to forgo our environment for the sake of growth, neither can actions for water be made a lesser priority compared to climate change and vice versa.

Navigating economic recovery post-pandemic and mitigating the impacts of climate change are highly dependent on the resilience of the water sector. A lot of countries are still grappling with economic recovery, whilst mitigation of disaster risks and adaptation efforts require billions of (dollars of the) development budget.

Malaysia would like to take this opportunity to also highlight that like other natural resources, water from the developing countries has benefitted the global economic growth and was unfairly exploited.

The global initiative for water security needs to acknowledge the connection between resilience and the investment needed, particularly addressing the global development inequity.

We have adopted a circular economy in waste and waste water management to support the nation's net-zero earliest by 2050 target . . .

We have also embedded more nature-based solutions that embody our long history of managing the abundance of water as a long-term strategy to adapt to climate change.

To further advance sustainability in our development, Malaysia has introduced the Water Sector Transformation 2040 Agenda. We envision the water sector to be an economic enabler and a dynamic growth engine to ensure water security for all . . .

The Water Action Agenda needs a framework that asserts water's role as the pillar that connects social, environmental, economic and cultural outcomes and be made a permanent agenda in the UN frameworks. For this purpose, a dedicated UN agency for water should be established.

It was truly a memorable experience, not something that I ever thought I would have the opportunity to do. But more than being a story to tell my grandchildren one day, it reinforced to me that water is increasingly important, not only to Malaysia but also the whole world. My call for the establishment of a UN water agency was picked up by the media.

As Fred Pearce, the popular environmental writer puts it, 'Water is the ultimate renewable resource.'[45] Transforming the water sector is crucial for building a more climate resilient Malaysia. If managed properly and responsibly, the value of water can be truly unlocked to contribute to human prosperity as well as a sustainable environment. There is a greater awareness of the need to let rivers remain in their natural state. This makes more sense than adding more concrete or straightening the courses of the waterways, which makes the communities living around the river more susceptible to floods. Recognizing the

---

[45] Pearce, Fred, *When the Rivers Run Dry: The Global Water Crisis and How to Solve It*. London: Granta Publications, 2019, p. 264.

risk of sea-level rise to coastal communities and resources, we must invest in adaptation, not merely mitigation.

Climate change essentially changes the water cycle. Overall, warmer weather means an increased rate of evaporation. This disruption leads to floods and droughts wreaking havoc on weather patterns across the world. Glaciers and ice sheets melt, rivers dry up, sea levels rise, and tropical storms become more intense. It affects how forests, mangroves, and oceans play their role as carbon sinks. Floods lead to loss of lives, spread of infectious diseases, and reduce the price of property. They also affect agriculture, thereby threatening food security. Droughts, on the other hand, increase the risk of wildfires. Drying rivers and reservoirs diminish the capacity of dams to generate electricity. Water scarcity will lead to greater demand for the resource, putting more pressure on water sources. Higher prices of clean water will put more pressure on the poorer and more marginalized segments of society. A warmer planet also affects the quality of water. With regards to water supply and waste water, we have seen the perils of privatization alongside the negative effects of not pricing water as a resource properly.

If we fail to manage our water resource, we will be a nation living in utter irony: we tend to have too much water yet not enough of it can be used due to our high consumption, massive leakages, and pollution.

Liveable Cities

Lord Mayor
We the bird community called a meeting
One fine clear morning
On the roof of the deserted Parliament building.

All sent their intellectuals to represent them,
All but the crows, for they were too busy
Mourning their loved ones, shot dead
And drifting down the River Klang.

Special guests came as observers,
A delegation of butterflies,
Involved in the issue.

Lord Mayor,
Though we had no hand in electing you
Since the franchise is not for the feathered
Still, we honoured you for your promise
Of a green city.

Alas, they have desecrated *the green* of nature
To worship *the green* of dollars
Since Kuala Lumpur's mud turned to concrete
We birds have been the silent sufferers
The late *Belatuk* was crushed under a felled tree

> *Merbuk* was conned by the name Padang Merbuk
> While he and his kind were cooped in cages.
>
> The *Pipit* delegation are protesting
> Against the insult in your proverb
> '*Pipit pekak makan berhujan*'
> (Deaf sparrows feed in the rain)
> *Pipit* and *Punai* both feel
> It's most improper of you to call
> Certain private parts of your anatomy
> By their names, when you well known
> Your *pipit* and *punai* can't fly
> (You have deflated our egos
> In the process of erecting yours).
>
> Lord Mayor,
> This letter requests that in your wisdom
> You will protect each branch, each root,
> Each leaf, each petal, each bower,
> For these have been our homes through the centuries,
> And it would also be for the good of man,
> His health and happiness, his peace of mind,
> To let nature and its myriad beauties bloom
> In the brilliant sun.
>
> —Translation of 'Surat dari Masyarakat
> Burung kepada Datuk Bandar'
> ('Letter from the Bird Community to the Mayor'),
> a poem by Malaysian National Laureate Usman Awang[46]

San Francisco has contributed much to music. Metallica, Santana, and the Grateful Dead are just a few of the bands that are from or

---

[46] Ishak, Solehah, et al., *Malaysian Literary Laureates: Selected Works*. Kuala Lumpur: Dewan Bahasa dan Pustaka, 1998, pp. 144–145.

were based in the city. In fact, journalist Herb Caen called the city 'Baghdad by the Bay' due to its diversity and cultural significance. But when talking about the *birdsongs* there, a curious observation was made. Since 1969, the songs of the white-crowned sparrows in San Francisco have been recorded for study. Similar to human beings, birds have 'dialects'—there is a geographical variation in birdsongs, and they evolve over time. How fast or slowly this happens depends on various factors. The frequency of the songs has increased over time, possibly adapting to the city's increased noise pollution. While dialects in the city's bushy countryside disappeared, the dialect that developed in the city not only survived but became dominant, replacing the disappearing dialects of the countryside.[47]

Throughout history, the impact of cities on humans has been clear. We are only now, however, realizing the impact of cities on animals, plants, and the environment. While climate change caused by carbon emissions is a relatively recent phenomenon, air pollution is a problem that predates the Industrial Revolution. Workshops and furnaces produced dirty and unhealthy air. The Greek physician, Hippocrates, advised travelling physicians to look at the cleanliness of a city's air and water. The use of fire from biomass for cooking, heating, and protection against mosquitoes led to lung diseases, which was recorded by various Roman scholars from over 2,000 years ago, as early as the dawn of the Common Era. At the same time, cases were heard in Roman courts about disputes over air pollution.[48]

About a thousand years after the birth of Hippocrates, Prophet Muhammad was born in Mecca. At that time, the city

---

[47] Luther, David, and Baptista, Luis, 'Urban Noise and the Cultural Evolution of Bird Songs', *Proceedings: Biological Sciences*, vol. 277, 1680. 2010, pp. 469–473, http://www.jstor.org/stable/40506142 [Accessed November 25, 2023].

[48] Mosley, Stephen, 'Environmental History of Air Pollution and Protection', in *The Basic Environmental History*, Agnoletti, Mauro and Simone Neri Serneri (eds.), Cham: Springer International Publishing, 2014, pp. 143–169.

was thriving with trade, attracting merchants from across Arabia and the Levant. Mecca was also already a city of pilgrimage, sustained by the Zamzam well. Yet, the infant Prophet was sent to the desert to be cared for by Halimah, a wet nurse. This was an Arab custom at that time to not only allow the young to learn the classical Arab tongue of the Bedouins but also enjoy the cleaner desert air outside the city.

Following the advent of the Industrial Revolution, London became notorious for its pollution. The city experienced the Great Stink of 1858 because of faeces, animal carcasses, rotting food, and industrial waste flowing into the Thames. This was followed by the Great Smog of 1952 due to exceptionally cold weather, leading to high usage of coal. Patients with respiratory ailments in the nineteenth century were prescribed a 'change of air'. Just like the Arabs more than a thousand years before them, they were instructed to go to a different locality, such as the seaside or mountains. This was an earlier form of the Malaysian balik kampung, ordered by the doctors.

In fact, it is argued that the exploitation of coal as a fossil fuel became a precursor not only to the Industrial Revolution but also to rapid urbanization.[49] High population density in cities often results in an increased use of other resources, leading to a higher concentration of air and water pollution. Rapid urbanization exacerbates this issue because worsening air quality, increased traffic, the creation of heat islands, a decrease in both green areas and permeable surfaces all contribute to potential environmental disasters and climate change.

At the same time today, unfettered capitalism is one of the major contributors to overconsumption, leading to pressure

---

[49] Mitchell, Timothy, *Carbon Democracy in the Age Oil: Political Power in the Age of Oil*. London: Verso, 2023, p. 15.

on natural resources, increased waste, and emissions. Living in cities, people inhabit high-density settlements. They go to school and work in close quarters. They live in the presence of and patronize major shopping malls. Resources are brought from far and wide to satisfy consumer demand. People are more exposed to advertisements. The Internet is widely available at ever faster speeds. As a result, the public lives beyond its means just to keep up. People end up drowning in debt. Consumerism creates a vicious cycle.

Rapid urbanization is also increasingly creating a disconnect between the city and nature. Green lungs are diminishing. As I wrote earlier, Malaysia has a tradition of balik kampung—returning to the village, particularly during the festive seasons and school holidays. This practice sees those who have migrated to cities and towns return to their kampungs to spend time with their parents and other family members. However, with ongoing urbanization, the next generation will more likely return to their parents' homes in cities and towns. Neither will they experience clear, pollution-free skies to stargaze any more nor will they see the origins of their food, such as rice or even where their favourite chicken nuggets come from. The 'kampung' in balik kampung will exist only in name.

I got a chance to go back to my mother's kampung in Kota Bharu (and even then, it was basically a kampung located in a town), but my son, Ilhan, has only experienced staying there overnight when he was a toddler, just after my grandmother passed away. He does not share my memory of sleeping over there, cousins squeezed in between one another, soaking in the unique smells and scents of the wooden house like I did. He will not experience having to use the outhouse or bathing from the freezing *kolah*, like I had to.

Malaysia is one of Southeast Asia's most urbanized countries with 75.1 per cent of our population living in cities and towns in 2020.[50] In 1970, the urbanization rate, at 33.5 per cent, was approximately half that. Globally, more than 50 per cent of the world's population lives in cities. But it was only in the 2000s that the global urban population overtook the rural. The higher carbon footprint of cities means that they account for 70 per cent of global carbon emissions.

In 1962, the professionals running Paris proposed a series of new roads cutting through the city centre as a solution to the city's traffic problems. This would have broken up the city and, as Malaysia's experience has taught us, would not actually have provided a sustainable solution to traffic jams. When Anne Hidalgo was elected mayor of the city in 2014, she made it her mission to prioritize pedestrians and cyclists instead of cars along with combating air pollution. This was initiated through the *Paris Respire* (Paris Breathes) campaign. This problem was particularly serious in France because the diesel prices were lower, and diesel is dirtier than petrol.[51] In 2024, the city residents voted to levy triple parking charges for SUVs, which tend to be owned by the rich, are more polluting, and potentially dangerous traffic hazards. Today, Parisians on scooters, rollerblades, and bicycles, along with pedestrians, are everywhere. The pedestrian and cyclist crossing lights are generally set to green. That clearly shows what Paris' priorities are. Compared to 1990, journeys made by car as a share of mobility in the city has dropped by half.

---

[50] Department of Statistics Malaysia, 'Key Findings of Population and Housing Census of Malaysia 2020: Urban and Rural', December 23, 2022, https://v1.dosm.gov.my/v1/index.php?r=column/ctheme&menu_id=L0pheU43NWJwRWVSZklWdzQ4TlhUUT09&bul_id=ZFRzTG9ubTkveFR4YUY2OXdNNk1GZz09 [Accessed March 4, 2024].

[51] Generally, in the past, diesel was taxed lower than petrol in Europe because, despite higher pollution, it is a more efficient fuel. Diesel engines can get better mileage, thus release less carbon dioxide, compared to petrol. But diesel engines produce more toxic nitrogen oxides and harmful particulate matter.

Even in the car-centric city of Kuala Lumpur, the Car Free Morning programme has been a tremendous success. When it was introduced in 2013, it was controversial, as businesses worried about losing customers and residents complained about the inconvenience caused by closed roads. Hell hath no fury like a Kuala Lumpur motorist scorned. But the programme only runs from 7.00 a.m. to 9.00 am on the first and third Sundays of each month, when most people are still asleep or at most loitering around the house. Today, the frequency has been increased to *every* Sunday. Essentially, you have a choice to go along the designated route, up to seven kilometres in the city's Golden Triangle walking, running, cycling, or using micromobility devices. Cars are obviously a no go. 5,000 people attend this Car Free Morning every week. I recall when I was doing a short run as the guest of honour for Bursa Malaysia—the Malaysian stock exchange—near the Car Free Morning route; we had random tourists from the latter event joining us just for the fun of it.

Conversely, the problems with cities—the high density and income that drives high consumption—are also their key advantages. Cities play a central role with regards to dealing with the triple planetary challenge to our climate, environment, and even biodiversity. Addressing the environmental footprint of cities will make a lot of difference in building climate resilience and having a sustainable environment.

We see climate targets being set at a global level including the COP meetings along with national governments. But for change to take place on the ground, the involvement of local governments is crucial. Even with countries committing themselves to ambitious global targets, without a buy-in and execution on the ground, they will not meet those targets. At the end of the day, cities provide an opportunity for action because local governments have the power to do more when national governments face constraints. Cities can pursue innovation and proof of concepts at a smaller scale for the national authorities to emulate. At times, being in the Federal Government feels like you have a huge lever to pull but getting things to actually move requires going through a long and slow process, with too many pulleys, wheels, and strings that need to move the right way for things to progress, let alone show results. Conversely, being in the state and local governments is like having a small electric button to press, which makes things happen instantaneously.

Cleaning Our Air

Imagine a child breathing in around thirty cigarettes a day. Yes, one and a half packets consumed by a child! That is what every Delhi boy and girl does if they breathe the city's air during the winter smog. Every year, when October arrives in Delhi, the city's population of over 33 million braces for the noxious and thick air pollution. With winter comes less rain and wind, worsening the air quality. Delhi ranks highest among cities suffering from this public health emergency. Satellite photos record a giant brown cloud hovering over Pakistan all the way to the northern part of the Indian Ocean, popularly dubbed as the 'Asian brown cloud'.

Much of the air pollution in this area happens because of open stubble burning by farmers. The use of firewood for cooking and heating homes also contributes to the problem (I earlier touched upon air pollution in the ancient world from the same sources). Then, as the Hindu celebration of Deepavali or Diwali takes place between the end of October to early November, things are made worse by the firecrackers burnt for the festivities. Following these celebrations, there is a 30 per cent spike in cases of respiratory diseases, particularly among children. Reportedly, a third of the city's children are asthmatic. A hospital in Delhi opened a pollution outpatient department, which not only includes ear, nose, and throat specialists but also psychiatrists. This is because many children who are not able to play outside due to the pollution suffer from mood swings and depression.[52]

Schools are forced to close, and construction works are halted. Firecrackers have been legally banned, but many people still use them. Residents who have the means to do so move out to the mountains, following age-old tradition, or at least purchase air purifiers. But, of course, not everyone has these options. Inspired by the concept of air purifiers, the government installed smog towers. These are essentially large-scale air purifiers similar to the ones used in homes and offices indoors. Yet other than giving a semblance of doing something, they do not work effectively outdoors.

The authorities distributed millions of face masks. Fossil fuel vehicles, open stubble burning, firewood usage, and firecrackers all release particulate matter—tiny particles of liquid or solid particles suspended in the air. Due to its small size, it can be inhaled, bypassing our body's natural filters, and settle in our lungs. Some particles may even reach our brain through our bloodstream, and they can potentially cause cancers. Moreover, particulates can also influence the climate, although their exact impact is difficult to predict.

In addition, as we were taught in school, nitrogen is by far the most abundant element in the atmosphere. When fossil fuels burn, they produce nitrogen oxides, in addition to particulate matter. This not only makes it worse for those with existing asthmatic

---

[52] Dhillon, Amrit, "'The Complete Murder of Our Young': India Counts Cost of Another Polluting Diwali on a Generation of Children', *The Guardian*, November 17, 2023, https://www.theguardian.com/global-development/2023/nov/17/the-complete-of-our-young-india-counts-cost-of-another-polluting-diwali-on-a-generation-of-children [Accessed November 17, 2023].

conditions but studies have also shown that it contributes to the *development* of asthma. When nitrogen oxides react with other components, it can harm our lungs, potentially causing lung cancer as well as heart problems and stroke.

Then, there is China. In the past, it was known as a 'bicycle economy' because that was a popular mode of transport among its citizens when the economy was closed until the 1980s. This was part and parcel of the 'work unit' or *danwei* socialist system under which citizens were provided lifetime employment and a social safety net, including the one-child policy. Everything was designed to be near the work unit—whether homes or schools—and thus, travelling by bicycle became not only possible but practical too.[53] However, as the economy grew, cars began to replace bicycles, releasing fumes in the air. Bicycle lanes were replaced with bigger roads to cater to cars. At the same time, the growing demand for electricity was increasingly met by coal. Factories sprouted up across the country, further contributing to pollution.

The government began to realize that something needed to be done. This became more urgent in the lead up to the 2008 Beijing Olympics. Coal power stations and factories were temporarily closed. One year before the event, the government appointed automotive expert Wan Gang as the minister of science and technology. Several electric buses and hybrid electric cars were made for the event. These buses operated in the central Olympic Park. A battery swapping technology was developed for electric vehicles.

---

[53] Zhang, Jun, 'A History of Bicycle Mobility in Urban China: Infrastructure, Economy, and Urban Planning (Part 1 of 2)', *MoLab Inventory of Mobilities and Socioeconomic Changes*, Department 'Anthropology of Economic Experimentation'. Halle/Saale: Max Planck Institute for Social Anthropology, April 2022, https://www.eth.mpg.de/molab-inventory/mobility-infrastructure/history-of-bicycle-mobility-in-urban-China-part-one [Accessed January 10, 2024].

After the 2008 Olympics, however, the smog returned to Beijing. In the following year, China became the world's largest market for cars, overtaking the US. Wan Gang managed to get approval to consider all types of electric vehicles, whether plug-in hybrids, fuel-cell cars, or pure battery-powered cars. Various subsidies and tax cuts were granted to grow the electric vehicle market.

This was complemented by the massive upgrade of China's railway system in 1997. The country today possesses the world's longest network of high-speed railway in the world—two-thirds of the total length of the world's high-speed railway.

The government was feeling the heat. Even though China is run as a one-party Communist state, environmental protests frequently broke out in the country, especially in the 2010s. Some, led by young students, even attracted up to tens of thousands of participants and successfully stopped environmentally harmful projects from being completed. At that time, the ratio of people killed by air pollution in China and India was three times more than in the West.

London has long experienced air pollution, dating back to the Industrial Revolution. William Blake wrote of the coal-powered, 'dark satanic mills' of the time. It is estimated that emissions in London were at least fifty times higher before contemporary laws were put in place.[54] The East End was particularly affected due to the high number of homes and factories there. As late as 1952, the Great Smog of London resulted in the death of at least 4,000 people and even cows in Smithfield due to air pollution. This was made worse by the colder-than-average winter conditions, causing many Londoners to burn vast amounts of coal for warmth.

---

[54] Air Quality Expert Group, 'Particulate Matter in the United Kingdom', *DEFRA*, 2005, https://uk-air.defra.gov.uk/library/assets/documents/reports/aqeg/Particulate_Matter_in_The_UK_2005.pdf [Accessed December 26, 2023].

The weather conditions also meant that the smoke could not rise. The fatal impact of the Great Smog led to the introduction of the Clean Air acts of 1956 and 1968, which effectively put an end to the severe smog of the past.

During my first year of studying in London in the early 2000s, the congestion charge was introduced in the city. This is a form of road pricing, just as tolls that we are familiar with on expressways on Peninsular Malaysia. While tolls are used to finance the construction of the infrastructure, congestion charges are imposed to internalize the costs of air pollution generated from the road users. Presently, these costs are externalized, adding to huge costs for the government and public in healthcare and reduced productivity. When the maverick Ken Livingstone contested for the first London mayoral election in 2000, he introduced congestion charge as part of his manifesto. It was implemented in 2003 for the London Inner Ring Road during peak hours.

Livingstone was replaced by Boris Johnson as Mayor of London from 2008 to 2016 (Johnson later became prime minister of the UK from 2019 to 2022). After London air pollution levels went above the EU limits in his second year as mayor, Johnson announced the creation of ULEZ to cover the same areas as the congestion charge in 2015. This was then put into place by his successor from Labour, Sadiq Khan. The difference is that ULEZ allows free access to electric vehicles, plug-in hybrids, and other low-polluting vehicles. This policy is applied at all times. In 2021, ULEZ was expanded to Inner London and, two years later, to the entire Greater London area. Income from ULEZ is reinvested in public transport.

The policy only affects one in ten of the cars in outer London. In fact, newer petrol cars have not been affected, as fuel economy has improved in recent years. The city has launched a GBP110 million policy to scrap affected cars for those with low incomes.

Khan made cleaning up London's notorious air pollution his personal mission, as encapsulated in the title of his book *Breathe*. Studies indicate that the cost of air pollution to the UK's health system is *at least* between GBP1.4 to 3.7 billion a year while a third of London schools are near roads with illegal levels of air pollution.[55]

The benefits of the responses have been clear. The number of older and dirtier vehicles has been halved. In 2023, 95 per cent of the cars on the move in the city comply with ULEZ standards compared to 39 per cent in 2017. From 2019 to 2021, nitrogen dioxide pollution was reduced by almost 50 per cent. Between 2016 to 2021, air pollution in the capital has fallen five times faster than the rest of the UK.[56] A study by British businesses in 2020 estimated that the ULEZ expansion could prevent 600 deaths and save more than 1,000 days of hospitalization a year.[57] The study also found that a longer life expectancy, reduction of illnesses, and fewer sick days also mean higher wages for workers.[58]

Heat Islands

Located in the shadows of the famous Petronas Twin Towers is Kampung Baru, an exclusive Malay quarter of Kuala Lumpur. Kampung Baru is less than a kilometre away, not even ten

---

[55] London Councils, 'Demystifying Air Pollution in London', January 2018, https://www.londoncouncils.gov.uk/node/33257 [Accessed November 21, 2023].

[56] City Hall, 'The Ultra Low Emission Zone (ULEZ) for London', https://www.london.gov.uk/programmes-strategies/environment-and-climate-change/pollution-and-air-quality/ultra-low-emission-zone-ulez-london [Accessed January 13, 2024].

[57] CBI Economics, 'Breathing Life into London', *Clean Air Fund*, April 2021, p. 2, https://www.cleanairfund.org/resource/breathing-life-into-the-uk-economy-cbi-economics-2020/ [Accessed January 13, 2024].

[58] CBI Economics, 'Breathing life into the UK Economy', September 2020, p. 27, https://www.cbi.org.uk/media/5539/2020-09-cbi-economics-caf-report.pdf [Accessed January 13, 2024].

minutes by foot from the iconic landmark. Established in 1899, the Malay Agricultural Settlement, as it was originally named, was intended to allow Malays to grow paddy to supply rice for the tin miners in the fast-developing city. While the surrounding areas were transformed by development, Kampung Baru remained pretty much a traditional Malay village protected by its unique legal status. Traditional wooden houses on stilts survive, and coconut trees along the streets are still common sights, just as they are in any other kampung. Malay cakes and packed nasi lemaks are sold throughout Kampung Baru. Other restaurants sell nasi padang from Sumatera and daging singgang from Kelantan—my favourite dishes. The visual juxtaposition of the traditional Malay kampung with the gleaming Twin Towers (the world's tallest buildings at one time) against the city skyline creates a stunning photograph. The image not only speaks to the city's vibrant diversity but also highlights the disparities between the rich and the poor, the global and local.

A few years back, a study compared Kampung Baru with the office of the government department managing the Orang Asli in rural Gombak about 20 kilometres away.[59] At its peak, Kampung Baru was 6 degrees Celsius hotter than Gombak. The only times Kampung Baru was cooler was when it enjoyed rain.

Modern cities tend to create urban heat islands, where the microclimate is warmer than the surrounding rural areas. The light and heat from the sun reach urban and rural areas equally, but cities tend to be warmer than rural areas. Tall and dense buildings, which replace forests and greenery, block wind and trap heat. More solar radiation is absorbed through surfaces such as asphalt. There are more carbon-emitting vehicles that contribute to air

---

[59] Sheikh Ahmad, Zaki, et al., 'Analysis of Urban Morphological Effect on the Microclimate of the Urban Residential Area of Kampung Baru in Kuala Lumpur Using a Geospatial Approach', *Sustainability*, vol. 12, 18: 7301. September 2020, https://www.mdpi.com/2071-1050/12/18/7301 [Accessed October 16, 2023].

pollution and haze. Kuala Lumpur has a heat island effect ranging between 4.2 to 9.5 degrees Celsius.[60] Think City found a 1.64 degrees Celsius increase in the capital city's surface temperature from 1989 to 2019, the worst being in the heavily developed Bukit Bintang, Pudu, and north Kuala Lumpur (including Kampung Baru). It is suggested that since Kampung Baru is a generally low-lying built area surrounded by skyscrapers, the urban heat island effect is more pronounced there. Additionally, studies show that traffic jams aggravate heat islands.[61]

One important yet seemingly simple natural solution, which most of us have heard of, is simply planting more trees. Trees help by providing shade. At the same time, they release moisture when the water collected on their leaves and the soil around them evaporates. They also release water vapour through the process of transpiration. This is why trees are said to be nature's own air conditioners. The Think City study found also that park reserve areas, such as Kuala Lumpur Forest Eco Park and Taman Botani Perdana, as well as public spaces and gardens in Mahameru and the area around Petronas Twin Towers were cooler.[62]

Malaysia has an ambitious 100 million tree planting campaign introduced in 2021, before my time as minister. I was asked in Parliament about the status of the programme. I acknowledged its importance but cautioned against making such programmes a greenwashing exercise. That surprised a few of the civil servants. But I explained to them afterwards that the best way to fight

---

[60] Wang, Kai, Yasemin D. Aktas, Jenny Stocker, et al., 'Urban Heat Island Modelling of a Tropical City: Case of Kuala Lumpur', *Geoscience Letters*, vol. 6, 4. April 23, 2019, https://geoscienceletters.springeropen.com/articles/10.1186/s40562-019-0134-2 [Accessed March 7, 2024].

[61] 'KL Traffic Jams May Trigger Urban Heat Island, Say Experts', *The Star*, July 9, 2022, https://www.thestar.com.my/news/nation/2022/07/09/kl-traffic-jams-may-trigger-urban-heat-island-say-experts [Accessed October 16, 2023].

[62] 'Land Surface Temperature', Think City, 2021, https://thinkcity.com.my/work/land-surface-temperature/ [Accessed September 17, 2023].

climate change is to stop deforestation. Protecting our forests has a bigger impact than tree plantation, which is more uncertain. It goes without saying that tree plantation is crucial, but we need to consider the survival rate and the time it will take for these trees to absorb the carbon in the atmosphere. So, while we should continue planting trees, our primary focus should preventing the destruction of our pristine forests!

Green Buildings

Green buildings are a crucial component of a more liveable city. The key is that we can mitigate the negative impact on climate and environment, from the planning, construction, and operations aspect. In fact, there are opportunities where green buildings can positively impact their surroundings.

As NRECC was the merger of two former ministries, I had the opportunity to locate my office in the grander (but older) natural resources premises on the Putrajaya Boulevard, which has views all the way to the prime minister's department or the more modest environment and water ministry building occupying one block at the Parcel F complex on the outskirts of Putrajaya. If the prime minister looks out from his office onto the boulevard, he can see the natural resources building. So, I surprised many by choosing the latter. My reasoning was simple, Parcel F is a gold rated GBI office. The entire complex occupies 34 acres of land and, with a built-up space of 417,000 square metres, has passive solar shading, sun-shading fins and naturally ventilated open spaces. This means that the view from my office, from where one can see the Kuala Lumpur skyline on a clear day, is obstructed by the shades. But it keeps my office cool. Solar panels generate 519 kilowatt peak of power. Rainwater and greywater are harvested. After the Cabinet reshuffle, the NRES was fortunately able to stay in Parcel F.

The Perdana Putra building—which hosts the prime minister's department, including his office and the Cabinet meeting room—was retrofitted in 2014 to go even further, with platinum rated GBI, more than a decade after it was completed. Energy consumption was reduced by more than 30 per cent after the upgrade. The first government building to get the platinum GBI rating was the Menara Kerja Raya, the headquarters for the public works department (PWD). The design reduces solar heat gain and energy consumption and maximizes natural lighting. Greywater from within the building, along with water from rainwater harvesting are used for toilet flushing and watering the landscape.

Buildings are significant consumers of electricity and water, and major contributors to carbon emissions. Green buildings are a low-hanging fruit that make the consumption of resources more efficient while reducing the emissions of greenhouse gases. Even simple steps, like correctly positioning windows relative to the sun, and planting trees in strategic locations, can greatly improve lighting and cooling in hot and humid countries like Malaysia. These changes also enhance the mental health and productivity of residents and employees. Designing buildings with public transportation and low-carbon mobility in mind is another important factor for green buildings.

Much Ado about Batik

In December 2022, shortly after my appointment as minister, I gave a media interview in which I noted how cold government buildings could be, as well as the impractical dress code for civil servants.[63] During our first Cabinet meeting, the men all wore suits and ties as per the dress code, except for the late Salahuddin Ayub, the minister for domestic trade and costs of living, who sat

---

[63] Vengadesan, Martin and Low Choon Chyuan 'Cold Govt Buildings, Thick Clothing - Minister Wants Mindset Change', *Malaysiakini*, December 31, 2022, https://www.malaysiakini.com/news/649922 [Accessed August 22, 2023].

next to me and had forgotten his tie.[64] Trying to justify himself, Salahuddin said he was certain that the rebel in Anwar Ibrahim would mean the Prime Minister would not wear a tie either. True enough, to my surprise, Anwar walked in, tie-less.

'I thought of relaxing the dress code,' Anwar said, smiling and taking his seat. We all took our ties off, taking the Prime Minister's cue. The MCKK Old Boys wear our old-school ties on Wednesdays, a tradition that, one rumour has it, started when Old Boy Abdul Razak Hussein was the prime minister and many of the ministers were fellow Old Boys. As the traditional day for Cabinet meetings took place on Wednesday (a practice that continues till today), that was the day chosen for us to wear our old college ties. Anwar, an Old Boy, has appointed two other Old Boys as ministers in the Cabinet other than me. But this time around, we won't be showing off our ties. The civil servants, however, were still in a bind, as they were bound by government circulars and at that point in time, it had not been changed yet.

During the first Parliament sitting for the Unity Government, again, Anwar took the lead by saying that ties were no longer required in the lower house. The Speaker ruled that this would be allowed by virtue of the prime minister being the leader of the house. Again, another conundrum emerged as the Opposition said unless the change was codified in the standing orders of Parliament, they will stick to the standing orders of lounge suit or national dress. So, we faced a situation where the Opposition, consisting of an Islamist party, whose leaders and some of its MPs tend to wear turbans and robes, and a Malay nationalist party, whose political stance was the primacy of being Malay and traditional, took a stand to defend the imposed dress code we inherited from our former colonial masters.

---

[64] The hardworking and dedicated Salahuddin passed away due to a brain haemorrhage just about eight months later.

I thought the situation offered us a chance to make some changes to the government dress code, which was very formal and colonial, often requiring a jacket and tie. Turns out, it was not easy to change certain things in the civil service. There was, many years back, a ruling mandating government officers to wear batik twice a month. But it did not take off. During Prime Minister Abdullah Ahmad Badawi's time, in 2008, the government mandated Thursdays as batik day. But this did not apply in Parliament, and my colleague, the then Permatang Pauh MP and KEADILAN Vice President Nurul Izzah Anwar, fought a long battle (when we were in government from 2018–2020) to convince the Speaker to allow batik on Thursdays. It was finally adopted in 2021, when the tourism, arts, and culture minister of the time, Nancy Shukri, met Speaker Azhar Azizan Harun.

After my interview, I wanted to push to allow civil servants to wear batik every day. I must admit, I was a late convert. But by 2018, batik made a comeback among the younger generation of Malaysian men (thus, jumping on the bandwagon, I guess, helped me to look young!). Around this time, cotton prints were enjoying a revival as opposed to the hand-drawn silk batik that had been typical of Malaysian batik for so long. It has started appearing in fancy shopping malls and while long-sleeve batik shirts are worn for formal occasions, the short-sleeve version is worn in more relaxed situations.

Finally, I decided to give it a try at a Cabinet meeting in early August 2023, when I presented a paper on the upcoming EECA bill in Parliament. In 2014, there was a government circular on energy conservation that set the temperature of government buildings and premises to be not less than 24 degrees Celsius. In my experience, until 2023, many premises did not follow that circular. There was not much discussion on the energy efficiency legislation per se but a lot on having batik as an option every day. There were some concerns about the cost of Malaysian batik. Some were concerned about it not being practical when meeting overseas visitors. But Nancy, now minister of women, family, and community development, argued in support of the change. Armizan Mohd Ali, a minister from Sabah, said the state had promoted its own ethnic batik motifs and also supported the move.

The Cabinet decided to proceed with the suggestion. Though it may seem like a small change, batik is suited to our climate. Implementing this change not only boosts the local batik industry but also fosters a distinct Malaysian identity. The decision to allow Malaysian civil servants to wear batik every day made headlines nationally.

Some questioned why I emphasized so much on this policy when the promotion of batik should fall under the culture portfolio. There were comments online questioning why a minister for climate change should be taking a stand on this matter.

The critics should look at the example of the Japanese Environment Minister Yuriko Koike who launched the Cool Biz campaign in 2005 to promote a no-tie, no-jacket dress code and reduce the use of air conditioning during summer months of June to September. After the meltdown of the Fukushima Nuclear Power Plant six years later, the campaign was extended from May to October. In 2022, Spanish Prime Minister Pedro Sanchez took off his tie in a press conference and asked public and private employees to stop wearing ties to save energy, as Europe had to cope with rising energy prices following the Russia–Ukraine war. Even the UK's conservative and staid House of Commons, which is a model for Malaysia's Parliament, allowed MPs to take off their jackets. Sometimes, there is a tendency for Malaysians to be even more British than the British!

Separately, our education ministry relaxed the school uniforms requirements due to the heatwave in 2023. Just like our civil service, our government schools have dress codes that are a legacy of the colonial era. While government schools do not have air conditioned classrooms, male students are all expected to be in shirts and often wear ties along with a jacket to boot! Changes were difficult, even changing the colour of school shoes from white to black proved to be a huge controversy at one point. Thus, the education ministry, in May 2023, allowed teachers and students to wear sports attire to school every day. The health, welfare, and safety of the teachers and students were placed as important priorities for the ministry in making this important decision. This was allowed until the end of 2023. For 2024, the government mandated wearing sports attire twice a week, still more than what was practised prior to 2023.

Whether it is to reduce air conditioner energy consumption or to deal with increased heat without air conditioners, re-evaluating our dress codes can go a long way. There are certainly occasions where protocol dictates formal attire being necessary, and I'm not disputing this; looking dapper in a full suit is fine. But we should be more flexible and practical in exploring alternatives to the rigid and outdated everyday dress code. Nowadays, I will try to wear long-sleeved batik to all possible official events. If I am primarily in my office or attending events as the chief guest, I even opt for short-sleeve shirts similar to what some corporates are doing in Malaysia.

During a green technology event, a group of ministers I was a part of who attended its opening were given batik shirts made from recycled plastic made by Kloth, a fabric recycling movement in Malaysia. That was truly a conversation starter, though it was a bit warm for the Malaysian weather, something the makers should consider in their future batik production! At the point I am writing this, though, Parliament still has not allowed MPs the choice of wearing batik every day.

Urban Forests

Urban forests play an important role in liveable cities. They lower the temperature, thus reducing the energy costs needed for air conditioning. Simultaneously, urban forests provide recreational areas that help keep city dwellers physically and mentally healthy, absorb runoff water to reduce flash floods, reduce sound pollution, and obviously absorb carbon.

While many recognize the iconic Christ the Redeemer Statue in Rio de Janeiro, not many know that it is situated on the Corcovado Mountain in the Tijuca National Park. This is the world's largest urban forest. In the nineteenth century, Rio de Janeiro faced water supply problems and flash floods. The Amazon rainforests had been replaced by coffee plantations, and when the plantations were abandoned, the slopes were left barren and exposed. This, in turn, affected the water springs supplying the city and farms. The government decided to implement a reforestation programme executed by five black men and one black woman—all slaves—Eleuterio, Constantino, Manuel, Mateus, Leopoldo, and Maria. 100,000 trees were planted between 1861 to 1887. Today, it contains many rare species, and its average temperature can be lower by up to 9 degrees Celsius than the temperature of the surrounding Rio de Janeiro urban area.

Back in Malaysia, Kuala Lumpur was historically part of Selangor until the Federal Government decided to carve it out as a Federal Territory in 1974. As part of this process of city planning for residential and recreational purposes, the government acquired three rubber estates: Hawthornden Rubber Estate, Bukit Jalil Estate, and Bukit Kiara Estate.

As mentioned earlier, one of the 'green' issues I had to confront as the Setiawangsa MP was Bukit Dinding, which originated in the Hawthornden Estate. A hill separating the suburb of Setiawangsa and Wangsa Maju, it was the site of the Battle of Bukit Dinding during the Selangor Civil War of 1870, which led to the British intervening in the state. In 1888, Bukit Dinding became part of the Hawthornden Estate. Indeed, much of Setiawangsa and Wangsa Maju were developed from the estate after it was acquired by the government in the 1980s and 1990s to provide more housing to the people of Kuala Lumpur.

Much of the development started from the lowest parts of Setiawangsa and Wangsa Maju, and then it slowly inched upwards as the two neighbourhoods grew over time. Part of the top of the hill is owned by a developer and has been zoned for residential purposes. It is yet to be developed till today. The peak of the hill—rising nearly 300 metres above sea level, making it among the highest peaks in the city—is government land and Telekom Malaysia (TM) has a telecommunication tower up there. The rest is a 50-acre recreational park. It attracts runners, hikers, and mountain bikers. The service road to the telecommunication tower is an attraction for leisure climbers and those wanting to bring their bicycles all the way up. As an Opposition MP, I raised the viability of the proposed project, as the technical assessments were done more than five years before the new proposal was submitted and before the density and height of the project were increased.

Since taking over as minister in December 2022, I have been exploring ways to gazette the urban forest in the recreational park as a forest reserve. The authorities also instructed the developer to conduct a new EIA. I have also collaborated with the local community in improving the situation for hikers and bikers on the hill.

The Bukit Kiara Federal Park, on the other hand, was part of the Bukit Kiara Estate. Unlike Hawthornden, the government acquired this estate for a public purpose. Part of it was proposed to be the 650-acre Kuala Lumpur Botanical Gardens. A nursery to provide seeds and trees to the garden was also established in Bukit Kiara. A top American landscape architecture firm was hired for the purpose. The former estate workers were housed in longhouses and were promised proper housing while they worked in the nursery. However, the gardens did not become a reality, and major parts of the land were developed by both public and private bodies. The rest of the land was naturally reforested over the years and a community movement to preserve it emerged. The Cabinet announced the creation of almost 467 acres public park in 2007. Nevertheless, 170 acres were compromised by existing commercial deals and another 60 acres could not be salvaged. In 2005, a regular hiker reported that the fireflies, which were commonly seen in the area, could no longer be found. Habitat loss, the use of insecticide, as well as urban light pollution are major causes affecting the numbers of fireflies.

Twelve years later, a hiker spotted fireflies again in Bukit Kiara. In fact, you can find the world's largest female firefly, along with six other species, in the urban forest in the middle of Greater Kuala Lumpur! In 2021, a citizen science project was initiated with various other organizations, involving members of the public as young as five years old![65] Today, Bukit Kiara includes a popular neighbourhood park (one of my favourite venues for a casual jog) as well as an arboretum. The authorities have replanted the arboretum with a variety of rainforest species to enrich the former rubber plantation.

Sponge Cities

The 'sponge city' concept was first developed by Kongjian Yu. He was born to a Chinese farming family in a small village called Dongyu in Zheijang, China. Growing up, he was tasked with handling the family buffalo and bringing it to the willowy marshes. As the country opened up its economy and modernized, new irrigation projects changed the course of traditional streams. Communities relied on massive water reservoirs. In cities, concrete jungles replaced permeable surfaces, such as urban forests or parks. As mentioned before, this meant that water discharge

---

[65] The project was initiated by the Friends of Bukit Kiara in collaboration with the Urban Biodiversity Initiative and the FRIM.

into grey infrastructure increased as the amount absorbed into the ground or by the greenery was reduced. The concept was to centralize runoff water flowing at high speed in mega concrete infrastructures. In highly developed cities, as little as a fifth of the rainwater that falls is absorbed into the ground.

Drawing on the Chinese concept that 'water is treasure', Yu developed the sponge city concept. It can create a cooler microclimate, slow down runoff water, and improve the quality of water that finally ends up in rivers and lakes. Wetlands are protected, not discarded. They can filter polluted water going into rivers and lakes. It is not simple to put into practice and can take many years, but the impact on cities is immense.

In July 2012, Beijing suffered a massive flash flood. Within a day, almost 60,000 people had to be moved out of the city while seventy-nine people died. Damages were calculated to be around US $1.6 billion. In one area, the water reportedly rose 1.3 metres in merely ten minutes!

The government took note of the floods and President Xi Jinping launched the sponge city initiative, with thirty projects launched by 2016, which were completed four years later. Each project covers a minimum of 13 square kilometres and aims to capture 70 per cent of the rainfall. Some of the projects include roads with permeable pavements, lowered gardens, green rooftops, canals as well as ponds.

As mentioned above, the folly of modern development making cities less climate resilient and more flood-prone is apparent in Jakarta. Sunda Kelapa, the old name for Jakarta's port, was a thriving hub from the thirteenth to sixteenth centuries. For a long time, mangroves protected the city from the tides. But when Java was colonized by the Dutch, who are famous for their flood mitigation measures, canals were built to deal with floods while the mangroves were cleared. As a result, the canals prevented eroded sediment settling in the rivers and replenishing the soil in the city, resulting in a sinking city.

Following a major flood in 2007, the provincial authorities required a minimum of 30 per cent of the city's area to be reserved as green space. At that time, only a tenth of Jakarta was green. Other than the aesthetic and recreational value, it also functions as a sponge to absorb the rainwater and recharge the underwater aquifers. This is to complement the existing grey infrastructure and traditional methods of dealing with floods: building dykes, dredging the river, and clearing the squatter areas.

Putrajaya, the administrative capital of Malaysia, was designed with many of the same principles of a sponge city, despite being developed earlier. Created as a greener and more organized alternative to Kuala Lumpur, almost 40 per cent of the city is designated as green space. The city is built around an artificial lake designed to cool the surrounding area. A wetland system has been constructed with various aquatic plants to filter the water flowing into the lake and slow down the runoff. This system attracts various migratory and local birds as well as other wildlife. There are also swale drains, which, in contrast to typical concrete drains, blend with the surroundings and are covered with vegetation.

Of course, relying on sponge cities and nature-based solutions *alone* may not be sufficient, especially when weather patterns become more extreme because of climate change. The city of Zhengzhou in Henan Province has been at the frontline of implementing sponge city principles. At the peak of the floods in July 2021, the city received 200 millimetres of rain in *an hour*, whereas the sponge city infrastructure could only handle the same amount of rain in *a day*![66]

In Malaysia, the Low Carbon Cities Framework (LCCF) was introduced in 2011. It crowdsourced the innovation of best

---

[66] Stanway, David, 'What Are China's "Sponge Cities" and Why Aren't They Stopping Floods?', *Reuters*, August 10, 2023, https://www.reuters.com/world/china/what-are-chinas-sponge-cities-why-arent-they-stopping-floods-2023-08-10/ [Accessed February 6, 2024].

practices from cities across the country in implementing low carbon policies and programmes. As part of this framework, first, a baseline is established by measuring and quantifying the carbon emissions for individual cities. Then, the cities are guided on strategies to reduce their carbon emissions. This then provides positive examples for other cities to emulate. Eventually, this leads to the creation of low carbon zones in major cities across the country. I have witnessed how leadership on the ground can make a difference—government departments, universities, and schools in Malacca come out tops in the low carbon awards that we hand out annually due to the state government's attention to this matter.

Five years after the introduction of LCCF, Kuala Lumpur joined the C40 Cities network. Today, almost 100 mayors from across the world come together as part of the network to collaborate on efforts to fight climate change. This has resulted in the Kuala Lumpur Climate Action Plan 2050 and the introduction of the Wangsa Maju Carbon Neutral Growth Centre programme. This area encompasses part of the Setiawangsa constituency that I represent. The idea is to leverage Wangsa Maju's well-planned township with new growth centres. It already has extensive public transport facilities—various bus services and an LRT network. The work has started—the installation of solar panels in schools and bus stops, extension of bicycle lanes, making school zones more pedestrian friendly, among other policies. In the future, floating solar panels will be installed on several water bodies in the area, an anaerobic digester that generates heat and electricity from organic waste will be set up, a self-sustaining eco-park to provide food for the local community will be developed, and low carbon challenges will be promoted in schools.

When I was appointed to the Cabinet, I got the idea of seeing climate and environmental policies in action at the local level and decided to come up with Ekosetiawangsa. This is a collaborative, bottom-up effort facilitated by NRES, the Kuala Lumpur City

Hall, and other agencies. It partly leverages the Carbon Neutral Growth Centre.

The examples highlighted above, if implemented consistently and adapted to local conditions, will hopefully help make Malaysian cities more liveable for all its people.

A Sustainable Environment

'Seperti enau dalam belukar, melepaskan pucuk masing-masing.'
('Like a sugar palm in the bush, growing its own shoots.')

—Malay proverb that means 'someone who only
thinks of themselves and not the rest'

For a long time, the modern capitalist economy has been fuelled by a desire for short-term profit. The impact on society and environment matters little if a company's stock price continues to rise, and they can regularly pay dividends to their shareholders. By delivering on this, the company's executives are rewarded with high salaries and fat bonuses. In economics, we are taught about negative externalities: costs caused by a producer, which are not financially incurred by them. Therefore, they only account for the direct costs reflected in their books, ignoring the costs imposed on the environment and society. This often leaves the public and government to shoulder these costs.

We can see an example of this in our reliance on fossil-fuel -powered mobility, a major driver of the Industrial Revolution since the nineteenth century. Similarly, this informs our traditional approach to managing rubbish. Sending waste to the landfills— often illegal ones to save costs even further—was the preferred way of dealing with it. One of the major contributors to waste is single-use plastics. Given its convenience and cheapness, we use it without thinking about its impact on the environment.

Then, there are the highly polluting industries that have been challenged to reduce the pollutants they produce. With the world heating up at alarming rates and the haze turning into a medical, economic, as well as diplomatic issue, these industries cannot take a business-as-usual approach. How can they clean up their act?

Without financial and economic costs attached to environmental degradation—even as people started to realize its immense impact on society—governments, corporations, and ultimately, much of the public decided to kick the can down the road. The attitude seemed to be that it is a problem to be dealt with by another minister, CEO, or our children for another day.

This was the backdrop to the birth of the modern environmental movement. A reaction against the use of chemicals in farming and water pollution caused by sewage and oil spills coincided with the counterculture phenomenon among the younger generation in the 1960s. Organic farming and recycling were spearheaded by these green pioneers. Many of the works of the people writing on the environment at that time were deemed as alarmist or rejected as doom and gloom naysayers. Many of them were dismissed out of hand.

Today, the markets have started to react by pricing in ESG concerns. This starts to give some indication of what the price of sustainability is. We have not properly valued resilience while ignoring fundamental risks.[67] This shift is positive, yet the environment needs to be the focal policy point. More attention needs to be given to climate adaptation—accepting that we can only mitigate so much—and adjust to the impact of climate change. This is necessary for Malaysia and the world to withstand what is coming for us.

---

[67] Carney, Mark, *Value(s): Climate, Credit, Covid and How We Focus on What Matters*. Revised and updated edition. London: William Collins, 2021, p. 230.

Green Mobility

As a result of air pollution plaguing its cities, China began to aggressively encourage more electric vehicles. More restrictions were placed on fossil fuel cars, making them more expensive. Car plates for fossil fuel vehicles were limited and had to go through a lottery whereas those for electric vehicles were more easily obtained. In September 2013, Beijing came up with a five-year plan to address air pollution. From a sale of 1,000 battery-powered cars and plug-in hybrids in 2011, the number shot up to nearly seven million eleven years later, making China the global leader in electric mobility. Today, a quarter of total cars sold there are electric, including Geely, owner of Volvo and Malaysian brand Proton. During Anwar Ibrahim's visit to China in 2023, he announced that the Chinese carmaker was planning a US $10 billion investment in Malaysia.

At the same time, our Prime Minister managed a coup by convincing the world's richest man, Elon Musk, to set up Tesla's Southeast Asian headquarters in the country. Tesla, the world's major electric vehicle and battery manufacturer, was also allowed to import two models into Malaysia. The Model Y sold in Malaysia has the second cheapest price globally. At the same time, Tesla started installing its superchargers in the country.

Growing up in the early 1990s, the big idea was Vision 2020. This was a plan for Malaysia to be a fully industrialized and developed nation by the year 2020. Our teachers would ask us to illustrate how the country would look like in 2020: often, these drawings would feature flying cars. Little did we know that instead of flying cars, electric vehicles would be the major revolution in mobility by 2020. Moreover, no one predicted a major pandemic shutting down the world that year.

Green mobility in Malaysia has been hampered by blanket petrol subsidies. Originally, the purpose of the subsidies was to keep fuel prices low. Since petrol consumption is a major expense for the public, subsidies were seen as a method to control inflation.

At present, petrol in Malaysia is graded based on RON.[68] Typically, RON 95 and RON 97 are sold in Malaysia. Thus, while diesel and RON 95 petrol prices are subsidized by the government, RON 97 fuel follows market prices. The idea is that the rich will choose the premium fuel. But those driving luxury cars willingly chose the inferior fuel simply due to the difference in price when global crude oil prices shot up. Fuel subsidies stall the adoption of electric vehicles because the difference is not significant enough to prompt a transition from fossil fuel vehicles, unlike in countries like UK or China where petrol is not subsidized.

In addition, subsidies result in a perverse effect where the government's limited funds (Malaysia has been running a budget deficit since 1998) are spent more on keeping petrol prices low for the rich instead of them aiding the middle classes and the poor. In 2024, the government began retargeting diesel subsidies. Instead of a wasteful blanket subsidy that fuelled smuggling across the borders, subsidies are now targeted for ordinary Malaysians through various cash transfers as well as specific subsidies for the logistics, agriculture and fisheries sectors.

Whenever I broach the topic of promoting electric vehicles, some environmental activists accuse me of ignoring the most ideal solution—public transport. For the record, I agree with Gustavo Petro, the then mayor of Bogota and later president of Colombia who said in 2013: 'A developed country is not a place where the poor have cars. It is where the rich use public transportation.'

That is why, in 2018, I protested Dr Mahathir Mohamad's cancellation of the MRT 3 project, which was supposed to pass through my constituency. Today, I am working hand in hand with the MRT Corporation as well as stakeholders to ensure

---

[68] A higher RON means a superior fuel that will perform better, particularly supercharged and turbocharged engines.

that the project can continue with a strong buy-in from the local community. Ironically, some of the people from the same group that accused me of ignoring public transport claimed that I was pushing expensive megaprojects at that time. The argument was that I was ignoring the cheaper solution—buses. Clearly, electric buses are the best option in terms of emission per capita, followed by intra-city rail. I understand the aversion to megaprojects due to our history, but without doubt, rail (which involves megaprojects) must be part of the solution.

In the 1980s, Dr Mahathir (you can see the pattern) prioritized the development of an automotive industry as well as a national car, a cornerstone of our foray into heavy industry and being a developed nation. This was laid down in 1991 as Vision 2020. With Proton and Perodua, Malaysia has two local car brands, making us the first to do so in Southeast Asia. By 2002, the former made Malaysia the eleventh country globally to make a car from scratch. Around the same time, with the launch of Proton as the first national car, the North–South Expressway was built—a backbone of the west coast of the Peninsula. The highway project was privatized. Soon, tolled highways grew in Malaysia to an extent that our highway system has been regarded as Asia's best highway system after Japan and Korea, as they are easily bankable and maintained due to the financial model. These factors, in addition to Malaysia being a petroleum producer, have made our cities very much centred on cars.

This has come at a cost. It has perpetuated the vicious cycle of a car-centric economy. It has hurt our intercity rail system that used to be the main mode of transport across Peninsular Malaysia, as it has failed to attract enough resources to compete. Until today, the Malayan Railway is owned by the government. I was the last batch of MCKK students to receive a third-class railway warrant (a practice inherited from the days of the British Empire where civil servants and students were given 'warrants' or vouchers charged to the government) to travel to and from Kuala Kangsar back to Kuala Lumpur. The diesel train took five hours to make the journey, whereas it took about three hours by highway. Prior to the completion of the highway, a journey by car would take about the same time as a journey by train.

The first two LRT and the monorail lines in the Klang Valley were poorly planned, and the ticketing system was initially not integrated, along with the rail commuter lines built on the existing rail network. Initially, the three lines were run by different private companies, with different systems. Even within the Klang Valley, corridors that could have been reserved for public transport were taken over by elevated highways that tore into the city.

Thus, I understand the cynicism of many Malaysians about any form of automobiles, including electric vehicles. But sustainable transportation needs to be looked at in its entirety: electric and hydrogen trucks and lorries in the logistical sector, electric buses and intra-city rail for urban public transport, electric and fuel-cell cars to complement it, and, not to forget, micromobility to provide that elusive last mile connection between rail and bus stops to the doorsteps of residents. Unfortunately, bicycles and micromobility are seen as a nuisance, not just by pedestrians but also by our car-centric culture. Our policies were designed with cars as the focus instead of other forms of transportation. While there are legitimate concerns with regards to road safety, it seems that, again, things have been seen more from the perspective of car-users.

At one time, a port operator in Malaysia even decided to introduce automated electric lorries to transport containers. This kills two birds with one stone: it deals with the high turnover of lorry drivers bored working in a monotonous job of driving all day long entirely inside the port area and reduces their emissions, which matters for many of their users.

The mini bus system flourished from the 1970s to 1990s. In their final years, the buses were painted in a shocking pink and white pattern. The mini buses packed as many passengers as possible, yet the conductors seems to always be able to issue a ticket to everyone on the bus. The drivers were always pushing the boundaries of the traffic laws.

Today, buses are enjoying a revival in Malaysia. In the Sunway suburb, an elevated BRT system has been operating since 2014. This public–private partnership project is the world's first all-electric system. A no-fare system, GoKL, was introduced in Kuala Lumpur's city centre in 2012. Initially, the routes were ones that were popular among tourists and foreign workers. When PH took over the Federal Government in 2018, the KL MPs, including myself, pushed for new routes to cover low-cost residential areas to make it relevant to our constituents. The system is now designed to take passengers from the city centre to residential areas. In Selangor, no-fare buses have been introduced by local councils starting 2015. In 2023, bus lanes began to be introduced in Kuala Lumpur, Malaysia, which was controversial because some drivers saw this as an inconvenience for cars. But these lanes help ensure the timeliness of buses, a crucial concern in Malaysia due to the notorious traffic jams. Again, it is important to have the political will to look beyond car users in order to have a more effective public transportation system.

Similarly, intercity rail that suffered because of the completion of highways is also beginning to challenge our car-centric culture on the West Coast of the Peninsula. This is because of the completion of the ETS in 2010. The fastest journey from Kuala Lumpur to Kuala Kangsar is less than three hours, akin to taking the highway. As the North–South Expressway gets badly congested during holidays and festivities, it is far more comfortable to take the train. The government is looking at reviving the dedicated Ipoh–Kuala Lumpur express to cater for the high demand for the route. A massive—and more controversial project—is the ECRL, connecting Port Klang on the west coast to Kota Bharu on the east coast. This is funded by China as part of the Belt and

Road Initiative. There are questions with regards to the project's viability. Unlike other intercity trains in Malaysia that run on meter gauge, it will run on the bigger standard gauge that will allow trains on ECRL to go faster. Thus, should I choose to ride the ECRL to Kota Bharu from Kuala Lumpur, my journey will be an estimated four hours—about half of what it takes by car. If I take the train to Kota Bharu today, I have to take the ETS south to Gemas, then change to the Jungle Railway all the way to Kelantan. The diesel-powered Jungle Railway is known for being scenic, but it takes a long time. All things considered, the same journey takes about fifteen hours now!

Looking at China's experience, electricity-powered trains not only overtook diesel trains in terms of length of network but, most importantly, their speed also increased from 120 to 200 kilometres per hour. The advent of high-speed rail contributed to more low carbon options for intercity mobility, while rail-based public transport catered to intra-city mobility. As we see in other countries, this has opened new commutes, as people choose to buy bigger and more comfortable homes further from the cities—the high-speed rail today makes it possible to work in a megapolis like Shanghai and live much further away. While, today, everyone takes it for granted that they can watch their favourite Korean dramas or cat videos on Instagram on their mobile phones during their commute, this practice actually started with the Japanese due to the time they spend on their commutes on their high-speed rail, the Shinkansen—a norm for Japanese workers.

Nevertheless, even in the most developed countries with highly developed, inter- and intra-city public transportation systems, electric vehicles have a role to play to reduce carbon emissions and promote a cleaner environment. After all, outside the Klang Valley, public transport is still very basic, and it is foolish to dismiss electric vehicles entirely. It can complement the various low-carbon mobility modes of transport.

There are also criticisms that electric vehicles are a luxury item, and the government should not be subsidizing the rich to own them. While I support targeted subsidies and advocate dismantling regressive and polluting blanket subsidies through lower electricity tariffs and fuel prices, I see no issue with initiating the adoption of electric vehicles through certain incentives. And yes, we need to develop the capacity to ensure proper and clean processing of batteries. Better recovery of resources in various electronic waste means that we need to mine less minerals from the ground, which, due to its extractive nature, is worse for the environment. Electric vehicle prices being set at the entry level (along with fuel prices following the market price) will be the gamechanger, as this will mean owning an electric vehicle makes perfect economic sense. Meanwhile, the government has started providing rebates for the purchase of electric motorcycles. Motorcycles have, of course, always been a popular choice as a cheap, fuel-efficient, and effective mode of transport that beats the traffic in Malaysia.

Hydrogen mobility, which utilizes fuel cells, also has huge potential. Fuel cells convert the chemical energy in hydrogen and oxygen into electricity. Its only byproducts are electricity, water, and heat. The technology has been around since the end of the nineteenth century. But its economics could not compete with petrol. However, today's demand to decarbonize energy and mobility has rekindled interest in hydrogen and fuel cells again. Hydrogen is seen as especially crucial for the logistics sector. Trucks fuelled by hydrogen can be filled with hydrogen faster than an electric truck being charged.

As mentioned above, Sarawak, which has access to huge amounts of hydroelectricity, has ambitious plans with regards to the hydrogen economy, including mobility. KUTS is an automated rapid transit system that utilizes dedicated lanes on the road and is powered by fuel cells. The premier and deputy premiers of

Sarawak also have been using the hydrogen fuel cell vehicle—the Toyota Mirai. The state already has a multi-fuel station that caters to vehicles powered by petroleum, diesel, electricity, and hydrogen, owned by Petroleum Sarawak (Petros).

As of 2023—the country of origin of Mirai—Japan boasts 168 hydrogen refuelling stations, heavily subsidized by the government. There are around 7,500 hydrogen-powered vehicles in the country, many of which are state-owned. Other than the Mirai, Honda has also produced the Clarity Fuel Cell. In 2015, there were only around 200 hydrogen vehicles in the country. Individuals get massive subsidies to buy hydrogen vehicles. Today, the cost for refuelling hydrogen vehicles is still prohibitive compared to electric vehicles. However, a Mirai only takes five minutes to be refuelled enough to go up to 650 kilometres. Using a Tesla Supercharger—one of the fastest electric charging systems in the world today—it still takes an electronic car fifteen minutes to be charged enough to go just over 300 kilometres.

Then, there is the shipping industry. Malaysia is a trading nation, and shipping is an important component of the country's economy. Peninsular Malaysia is situated on one side of the Straits of Malacca, which forms a vital artery of the Maritime Silk Road and is one of the busiest shipping routes in the world. At present, ships are primarily powered by dirty fuels, such as heavy fuel oil, marine gas oil, and marine diesel oil. These are among the dirtiest fossil fuels in existence. Methanol and hydrogen are among the fuels that can be considered as options for a greener future for the maritime industry, along with electric ships. A Malaysian company, Yinson GreenTech, is developing electric vessels.

If you look at the map of Scotland, you will see the Orkney Islands off its northeast coast. As you can imagine, its location means that the islands have significant wind energy. They are also exploiting wave and tidal energy. This allows them to rely entirely on renewable energy. In fact, even after exporting energy to the rest of the UK through the national grid, they have over 100 per cent excess energy. Orkney is using this excess power to manufacture green hydrogen. It is being used to power not only hydrogen vehicles in the local council but also vessels, including passenger ferries and even small passenger aircrafts.

As mentioned, Malaysia has introduced LCCF for local governments, universities, and states to compete on low-carbon strategies for buildings, transportation, infrastructure, and environment. This pushes cities and establishments to find ways to make the transition to lower carbon mobility.

Putting a Price on Carbon

In December 2022, I launched the BCX, the world's first sharia-compliant carbon exchange. The BCX allows companies to trade carbon credits to fulfil their climate objectives. The credits traded are certified by Verra, one of the popular standards used for carbon offsets. The exchange also plans to include carbon credits certified by Gold Standard, another popular standard, as well as RECs issued for clean energy generation. BCX is a voluntary carbon market with no restrictions on exporting the carbon credits to purchasers outside Malaysia. The first auctions on the BCX were conducted in March the following year, when 150,000 credits were sold.

Carbon credits are generated by scaling down or eliminating greenhouse gas emissions from the atmosphere. This can come from nature- or technology-based solutions. Nature-based solutions include improving agricultural practices to retain more carbon in the soil or avoiding deforestation to preserve the function of forests as a carbon sink. Technology-based solutions include renewable energy projects, collection of methane from farming or landfills, as well as energy efficiency measures.

Avoiding deforestation is something that can be implemented in Malaysia and fetches a premium in the market. This can be achieved by repairing degraded forests or expanding forest reserves. But those pursuing the projects must show that they need the funds generated from the carbon credits. This is what is called 'additionality'. If it can be shown that the projects would have happened anyway, then there is no additionality, and the credit has no value to offset carbon. For example, an existing forest reserve cannot generate carbon credits.

In 2021, the public was shocked over a carbon credit deal supposedly involving two million hectares of forest reserves in Sabah that will be protected from being developed for 100 years. The major criticism was that the indigenous people were not consulted. The UN has established that indigenous peoples have the right to consent to carbon credit deals in their territories through the FPIC principle. At the same time, the deal would have allowed a foreign partner incorporated in a tax haven to receive a 30 per cent share of the proceeds from the sale of carbon credits. There was also concern about the project's claim to involve permanent forest reserves, which begs the question of how additionality would be achieved.[69]

To be sure, there is a promising carbon credit project in Sabah—the Kuamut Rainforest Conservation Project—that is being implemented properly. It went into auction at the BCX in July 2024. This public–private partnership in Tongod and Kinabatangan districts involves over 80,000 hectares of forests that were being logged. The project received Verra certification and received triple gold rating with regards to climate, community, and biodiversity. Many buyers of carbon credits today want to go beyond carbon offsetting alone. The project aims to ensure that the forest is conserved and repaired to absorb carbon from

---

[69] Cannon, John, 'Critical Questions Remain as Carbon Credit Deal in Sabah Presses Forward', *Mongabay*, November 1, 2023, https://news.mongabay.com/2023/11/critical-questions-remain-as-carbon-credit-deal-in-sabah-presses-forward/ [Accessed February 9, 2024].

the atmosphere, to uplift the socio-economic standing of the indigenous communities living around the forest and to ensure that the tropical rainforest can not only survive but also thrive, protecting the plants and animals that inhabit it. The forest is home to elephants, *banteng* (a type of wild cattle), and orangutans. Those promoting Kuamut work with a credible local indigenous organization, the PACOS Trust. The land will be reclassified as a forest reserve upon receiving climate financing. It is estimated that it can reduce a total of over 15 million tonnes of carbon dioxide equivalent in thirty years.

The ICAO has introduced a carbon credit regime known as CORSIA. At present, airlines participate in the programme voluntarily as a means to mitigate carbon emissions from aviation, a high emission mode of travel. At the point of writing this, Malaysian airlines have had to resort to foreign carbon credit projects to fulfil their objectives under CORSIA.

There are criticisms of the effectiveness of carbon markets, and the risk that they are mere greenwashing. First, the argument goes that they do not provide sufficiently drastic measures for us to reduce carbon emissions. Carbon offsetting creates the impression that we can pursue our current lifestyle without making big changes as long as we offset the carbon that is produced. Second, it focuses on the choices of individuals and corporations rather than the political decisions needed for far-reaching changes. Third, there is a question mark about the integrity of carbon credits produced by avoiding deforestation as well as human rights issues emerging from those nature-based carbon projects. The bottom line is that the focus should be on directly reducing carbon emissions rather than trading them.

Nevertheless, it is important to note that reducing emissions alone, without protecting forests, a major form of carbon sinks, is not sufficient to avert what scientists consider the red lines of global warming. Companies that participate in voluntary carbon markets have reduced emissions reductions compared to the companies that don't. The same companies purchasing carbon credits tend to invest more in future carbon reductions.[70] There have been successful reforestation projects through carbon offsetting, including in Costa Rica and the Peruvian Amazon. Methodologies for carbon offsetting are being upgraded to improve their accuracy and credibility. So, while carbon markets alone have their limitations, they can still play a useful role in our fight against climate change.

There are other forms of carbon pricing based on the concept of internalizing the cost of carbon. Emissions trading is rooted in an effort to deal with air pollution and acid rain in the US, dating back to the Clean Air Act of 1977. The EU launched the Emissions Trading Scheme in 2005. This is the world's largest cap-and-trade programme. Essentially, companies are allocated emission caps or a carbon budget. They can trade the allocated permits for emissions. In other words, companies that produce fewer emissions than their budgets can sell their permits to companies that have exhausted their budgets. Over time, the total carbon budget is reduced. Although the Emissions Trading Scheme started with a low price (the general view is that for such a scheme to be effective, the price of emissions needs to be higher), research has shown that it has successfully reduced carbon emissions.[71]

Carbon taxes are imposed by the government to ensure that the cost of emitting carbon is reflected in prices of goods and

---

[70] CDP, 'CDP Climate Change Report 2016', October 2016, p. 34, https://cdn.cdp.net/cdp-production/cms/reports/documents/000/001/233/original/CEE-edition-climate-change-report-2016.PDF?1478599986 [Accessed February 9, 2024].

[71] Bayer, Patrick and Michaël Aklin, 'The European Union Emissions Trading System reduced CO2 emissions Despite Low Prices', *Proceedings of the National Academy of Sciences of the United States of America*, vol. 117, 16. 2020, https://www.ncbi.nlm.nih.gov/pmc/articles/PMC7183178/ [Accessed February 10, 2024].

services. They are simpler to administer compared to Emissions Trading Scheme. For corporations investing in carbon reduction measures, taxes provide certainty.[72] It can also cover a wider range of emissions. At a time when governments across the world face limited fiscal space and substantial debts, this is a way to broaden the tax-base. The Malaysian government, led by the Ministry of Finance and assisted by the NRES, has been in discussions with the World Bank to explore the possibility of implementing carbon taxes in the country. However, the issue of blanket fossil fuel subsidies for petroleum, diesel, and electricity must first be addressed. As explained earlier, we have started with electricity and are looking to introduce targeted subsidies for petroleum and diesel. The drawback for carbon tax is that studies show these taxes hit the poor the worst and, therefore, are regressive.

What to Do with the Waste?

Ask many Malaysians—and not just Muslims—and they will say the same thing: one of the most iconic features of the holy month of fasting in the country is the Ramadan bazaars. Springing up in cities and towns across the country in late afternoons, they cater to those looking for food to break their fasts in the evening. However, Malaysians of other faiths looking for unique fare that is only available during the month also visit them. The smell of grilled chicken and satay (or grilled beef fat in Kelantan), sweet *kuehs*, and comforting and colourful ice drinks are all part of these bazaars. As an elected representative, I always drop by as many Ramadan bazaars as possible because it is a good way to meet my voters. I often distribute dates, a traditional food for breaking fast. Some food sellers in Ramadan bazaars prepare too much food to garner more sales. In 2023, it was estimated that almost 48,000 kilogrammes of food—enough to feed 40,000 people—

---

[72] As opposed to ETS, where the emission reduction is defined but the price is left to the market.

was thrown away daily in Kuala Lumpur and Putrajaya. Every bazaar lot produces 9.5 kilogramme of food waste every day.

The affluent mostly break their fasts with expensive buffets in five-star hotels, again resulting in waste. Unfortunately, food waste during Ramadan and other festivities in Malaysia increases up to 20 per cent higher than usual. This is ironic, as Ramadan teaches Muslims to fast to control the self and to achieve God-consciousness. This does not even account for the widespread use of plastic bags and containers during this time, increasing the ecological footprint of the country during the holy month.

Waste management is a major environmental issue for Malaysia. In 2017, up to 80 per cent of the budget of local government went into solid waste management.[73] Every day, more than 38,000 metric tonnes of solid waste is produced at a municipal level. Most of this is food waste, followed by plastic and paper. In one week, Kuala Lumpur produces almost 18,000 metric tonnes of rubbish, which, if piled up, would stand as tall as the world-famous Petronas Twin Towers. Most of this waste ends up in one of the country's 165 landfills. The volume disposed is growing beyond the country's recycling rate. If current trends continue and no major policy changes are made, the country is expected to run out of space for landfills by 2050. Even when properly managed, landfills take about twenty years of recovery and restoration before they can be used for other purposes.

Rain and other forms of water that gather in landfills create leachate. If the landfill is not properly managed, the leachate pollutes groundwater. The rotting rubbish also produces methane and carbon dioxide. In my first year as a minister, I went to two illegal dumpsites in Selangor, dubbed the Twin Palms and Black Water Lake. There is another, rather scenic Blackwater Lake in

---

[73] Zainu, Zaipul Anwar and Ahmad Rahman Songip, 'Policies, Challenges and Strategies for Municipal Waste Management in Malaysia', *Journal of Science, Technology and Innovation Policy*, vol. 3, 1. June 2017, pp. 10–14.

Canada, whereas its Malaysian namesake is given the name due to the black, polluted water there. A huge swath of rubbish has been dumped in the lake, which was formerly a mining site. The cost to restore the lake to its pristine condition would be over RM40 million. It is located not far from the permanent forest reserves that span Selangor's border with Pahang.

When I was growing up, there was a big landfill in Kelana Jaya. It took in rubbish from three cities in the Klang Valley: Petaling Jaya, Shah Alam, and Subang Jaya. The landfill closed in 1996. Less than a decade after that, houses were developed there. When the purchasers renovated their homes, they found leachate and layers of decaying rubbish. The homes began to sink, metal turned green due to rust while the ground was contaminated by mercury. When another project was built there, a journalist saw lorries carrying decaying rubbish from the site, sending them not to proper sanitary landfills (which are located far away) but just close to a lake near homes and a school. I know this because, by then, I was the state assemblyperson for the area![74] The Department of Solid Waste Management does not actually allow the redevelopment of landfills into residential areas due to the unstable land conditions as well as polluted soil. Normally, such areas are turned into parks. But as the urban sprawl proliferates unabated, these pieces of land far from the city-centre become very valuable.

It is estimated that 10 per cent of food thrown out every day in Malaysia can still be eaten. Daily, more than 4,000 metric tonnes of edible food—enough to feed three million people with three daily meals!—is thrown away.[75] Meanwhile, supermarkets

---

[74] Tan Cheng Li, 'Mixed Views on Redevelopment of Landfills', *The Star*, August 16, 2011, https://www.thestar.com.my/lifestyle/features/2011/08/16/mixed-views-on-redevelopment-of-landfills [Accessed February 9, 2024].

[75] Audrey Dermawan, 'CAP: Crack the Whip on Food Waste', *New Straits Times*, August 29, 2022, https://www.nst.com.my/news/nation/2022/08/826643/cap-crack-whip-food-waste [Accessed October 2, 2023].

throw away up to one-fifth of 'ugly' fruits and vegetables simply because they are oddly shaped or have blemishes, although they are perfectly edible.

I met the people behind TUG, a Malaysian brand that turns 'ugly' fruits into gelato (and tasted it!). Established by Hailey Wong when she was nineteen, it was inspired by her participation in the Hult Prize, known as the Nobel Prize for students. This competition challenges youths to find solutions to the big social issues of the day. In 2021, the theme was 'Food for Good'. She read up on food waste, and while she was at the market, saw how fruit-sellers separate fruits based on their appearance. Taking time off from university, she started TUG by the end of the year.

While some environmental activists consistently remind us of the impact of meat eating (to be sure, eating less meat is definitely good for your health and the environment), let's not forget that food wastage is a bigger issue, and, in fact, a low-hanging fruit. Asking Malaysians to give up their beef *rendang, bak kut teh*, chicken curry, and *hinava* may be unrealistic. Instead, wouldn't it be better to appeal to what parents have been reminding their children since time immemorial? Do not waste food.

Managing waste cooking oil is another important environmental issue. While it is supposed to be disposed in a container in the garbage bin, often, as a liquid, it is discarded in the plumbing system, clogging up pipes and sewers. When the oil goes further into waterways, it spreads thinly on the surface, preventing oxygen from reaching animals and plants in the water.

The authorities require commercial food and beverage operators to have grease traps that collect cooking oil. While many install these traps to comply with licensing requirements, several operators refuse to use them properly due to the cost of disposing waste oil.

Nevertheless, today, there is a major demand for sustainable aviation fuel to reduce the carbon footprint of the aviation industry. This offers an interesting solution for used cooking oil. FatHopes Energy is a Malaysian company that is a major Southeast Asian biofuel player across the supply chain. From individual agents collecting used cooking oil in local neighbourhoods all the way to big contracts with fast food companies, FatHopes collects waste, verifies the integrity of the supply (crucial in the process to manufacture sustainable fuels), and turns it into biofuel feedstock. Some parent–teacher associations in schools and mosques collect used cooking oil to fund school and mosque activities. FatHopes works with oil and gas supermajor BP. Other alternatives, like the production of sustainable aviation fuel from algae, are being explored by Petronas and a Sarawak state government agency.

Recycling is a crucial aspect to reduce waste and complete the circular economy. However, in Malaysia, only one-third of recyclable waste is actually recycled. There has been a steady increase in the rate, but it still seems like a tall order to achieve the target of 40 per cent by 2025. This will require a lot of drastic measures. The conundrum is that many recycling companies do not have sufficient domestic trash to recycle and resort to importing foreign garbage for their operations. We must work on making domestic recycling more viable. This became worse when China banned the import of plastics inside its borders. It was estimated that, between 2018 to 2030, 111 million metric tonnes of waste needed to find an outlet, in light of China's new policy. The developed economies looked for a solution, and Malaysia became one of them. In short, we became a garbage bin for the world. While there were legitimate plastic recyclers who operated in compliance with our environmental policies, others processed contaminated plastic illegally, aided by the availability of cheap foreign labour.

Space is a major constraint for Singapore. Thus, all waste in the city-state that is not recycled is incinerated at WTE plants, where temperatures reach 1,000 degrees Celsius. Sweden became the envy of the world in 2005, when it banned combustible and organic rubbish being sent to landfills. Most of its waste was either recycled or incinerated in WTE plants. Only 1 per cent of their trash ends up in landfills, 52 per cent is converted to heat, and 47 per cent is recycled. The heat is turned into electricity or heats people's homes. In fact, Sweden was so successful that it had to import rubbish to make up for shortfall in feedstock to power its WTE plants. The UK, Ireland, Italy, and Norway *pay* Sweden to import trash from their countries. Awareness has been built through various measures. Pantamera, a civil society organization, has involved popular celebrities in recording songs to promote recycling. The policy in Sweden is to have a recycling station within one mile (1.6 kilometres) of every neighbourhood.

Nevertheless, WTE plants face their own challenges. Sweden's success lies in the fact that much of its waste goes to recycling and WTE plants. The toxic fly ash residue produced from incinerating the rubbish used to be released. But due to air pollution regulations, it is now captured and exported to Langøya island and Brevik in Norway to be buried in the ground. At the same time, Sweden's success in being able to process waste cheaply means that Norway struggles to find rubbish to fuel its own WTE plants. Sometimes, the lack of waste means that Norway has had to burn rubbish that could be recycled. It is ironic that waste is being exported to developing countries, when some developed countries still need waste for their own plants!

In Singapore, the fly ash residue from WTE plants is transported to the Semakau island. It is deposited in the middle of the island, surrounded by a rock bund to protect the sea surrounding Semakau. However, it is predicted that in 2035, Semakau will be full. A new WTE plant is expected to be needed in the city-state every seven to ten years with offshore landfills every thirty to thirty-five years. Due to the scarcity of land, this is unsustainable for Singapore. A WTE plant proposed in Malaysia is planning to use the fly ash residue in their cement for construction.

Without doubt, WTE plants with the latest emission control designs supported by strict government regulation must be part of the solution for waste management. Yes, reducing waste, recycling, and composting must be the priority but without the option of WTE plants, dirty and polluting landfills will continue to expand quickly.

The Problem with Plastics

> I recently purchased cut papaya fruits in Singapore, which came from a plastic bag, inside a plastic container that checkout staff subsequently tried to place inside another small plastic bag.[76]

This observation was made by Assaad Razzouk, a Lebanese–British clean-energy entrepreneur based in the city-state. Although we Malaysians have a tendency of feeling schadenfreude over any criticism of our neighbour down south, let's be honest about this being a common practice in our country as well. Razzouk focused on how other developing countries were faring better on this front. Rwanda has become one of the cleanest countries in Africa following its total ban on plastic bags in 2008. Bangladesh, too, has banned single-use plastics, as the government recognized the material as a contributor to the country's flooding problem.

I witnessed this first-hand in Malaysia. During a visit to a flash flood site in Klang, I saw rubbish, mostly plastic bottles, stuck at the culvert. For some time, we have been told about the problems of plastic waste, particularly single-use plastics. It takes 450 years or more for plastic bottles to biodegrade. In comparison, aluminium cans take eighty years, takeaway coffee cups thirty years, cardboard two months, and paper two to six weeks. A lot of these plastic materials, left behind in parks, near rivers, and

---

[76] Razzouk, Assaad, *Saving the Planet without the Bullsh*t: What They Don't Tell You About the Climate Crisis*. London: Atlantic Books, 2022.

on beaches litter land and waterways, ending up in the ocean as a danger to marine life.

Image 15: Having a bilateral with Dr Claudine Uwera, Rwandan minister of state for the environment, at the sidelines of the UN Environmental Assembly 2024 in Nairobi, Kenya. Rwanda is among the first countries in the world to ban single use plastics, making Kigali one of the cleanest cities in Africa.

In 2023, an image on VisualCapitalist.com attracted a lot of attention. It showed a sphere of floating rubbish on the ocean.

It was divided according to each country's contribution to ocean plastic pollution based on a 2021 study. The study listed Malaysia as the third biggest global contributor to ocean plastic waste. Malaysia generates almost a million metric tonnes of plastic waste a year and China 12 million metric tonnes more. However, 9 per cent of Malaysia's plastic waste reaches the ocean, compared to 0.6 per cent for China.[77] When the study came out, it did not garner much attention in the country. But when the graphic came out, it made news.

But it was during the Klang visit that I was struck by the enormous impact of plastic waste on the environment. The water flow at the culvert was obstructed by these plastic bottles. The durability of plastic allowed it to remain intact despite the massive amounts of water in the drain.

As part of our preparation for the 2023 monsoon, I visited the various flood retention ponds in Kuala Lumpur to assess our preparedness. The problem of plastic waste was particularly visible during my visit to the Jinjang River Flood Retention Ponds. These three ponds highlighted the problem of plastic waste for everyone to see. Plastic waste remained intact, clogging water flows to and from the ponds. The amount of water increasing during the rainy season was compounded by a reduction in the capacity of the ponds due to the amount of silt and waste flowing into them. All sorts of garbage—mostly from the Selayang Wholesale Market, the country's biggest wholesale mart—made its way to these ponds. A strong, fishy stench emanated from them, but it had nothing to do with fishes swimming in the pond. It was coming from the polystyrene ice boxes that stored fish from the market. Plastic and polystyrene waste is the most noticeable and

---

[77] Meijer, Lourens J. J., et al., 'More Than 1000 Rivers Account for 80% of Global Riverine Plastic Emissions into the Ocean', *Scientific Advances*, vol. 7, 18. April 30, 2021, https://www.science.org/doi/10.1126/sciadv.aaz5803 [Accessed September 17, 2023].

lasts longest due to its non-biodegradable nature. The DID has to clear this rubbish twice a week—a heavy cost for them. If they spent less on clearing the rubbish, they could spend more on desilting to maintain the capacity of the ponds.

Such plastic trash goes into rivers and ends up in the sea. Four-fifths of trash in the ocean comes from land, particularly from riverine and coastal cities. In 1997, oceanographer Charles Moore returned to California after finishing a sailing race from Los Angeles to Hawaii. In the middle of the Pacific Ocean, he caught sight of what is now known as the Great Pacific Garbage Patch—a huge collection of marine rubbish.

> I often struggle to find words that will communicate the vastness of the Pacific Ocean to people who have never been to sea [. . .] Yet as I gazed from the deck at the surface of what ought to have been a pristine ocean, I was confronted, as far as the eye could see, with the sight of plastic.
>
> It seemed unbelievable, but I never found a clear spot. In the week it took to cross the subtropical high, no matter what time of day I looked, plastic debris was floating everywhere: bottles, bottle caps, wrappers, fragments [. . . Oceanographer Curtis] Ebbesmeyer has estimated that the area, nearly covered with floating plastic debris, is roughly the size of Texas.[78]

That is more than double the size of Malaysia. The location is distant and remote, yet it brings together rubbish from both sides of the Pacific, particularly from Japan and California. It is where the warm waters from the southern part of the ocean meet with Arctic waters. As we all know, plastics do not decompose or degrade but, for many hundred years, break down into

---

[78] Moore, Charles, 'Trashed: Across the Pacific Ocean, Plastics, Plastics Everywhere', *Natural History*, November 2003, https://www.naturalhistorymag.com/htmlsite/1103/1103_feature.html [Accessed September 17, 2023].

microplastics smaller than ornamental pearls, mixed with fishing nets and other rubbish. In fact, microplastics can break down to even smaller particles, nanoplastics. Thus, these forms of plastics are pretty much part of the marine landscape now. The deepest trench in the ocean is the Mariana Trench in the Pacific that goes as deep as almost 11,000 metres. In an expedition down to the depths of the trench in 2019, an explorer discovered candy wrappers and plastic bags.

Image 16: Responding to questions from students and lecturers at a forum on plastics at Universiti Malaysia Sarawak

From the biggest whales to the smallest plankton, microplastics have been found in all creatures. As a result of being in the food chain, microplastics are present in humans. Babies, in fact, encounter microplastics even before being born, as the material has been found in their placentas.

The attraction of plastic is that it is cheap and durable. While plastics have been around since the 1800s and have made major contributions to cars, electrical products, and medical devices, it was only after the 1950s that it could be mass-produced, bringing down its price. It has been dubbed the miracle invention. Since the 1970s, single-use plastics—in the form of water bottles, bags, and straws—have become popular. Prior to that, the older generation brought their own bags to the market as well as packed rice and curries in tiffin carriers. Changing lifestyles—the growth of the fast-food industry and shopping malls—have contributed to increased use of disposable plastics. Single-use plastics are made to be used for a short period of time and then chucked as waste. Additionally, many plastic products have become cheap. In the past, people would find ways to repair and reuse toys, electrical products, and clothing (yes, the 3Rs of reduce, reuse, and recycle that are popularly cited today are, when you think about it, basically a call to emulate the lifestyles of the past eras). But now, it's easier to just chuck everything in a bin and buy it anew. Even the ubiquitous microbeads present in exfoliating facial cleansers are made from plastic (though there are biodegradable alternatives such as coffee grounds, jojoba beads, and bamboo powder). If one truly internalized the actual cost of plastic, you'll realize that it's really very, very expensive.

Let's not forget that 60 per cent of clothing is also made from plastic. This includes polyester, nylon, and acrylic. In our homes, polyester is in everything from our bed sheets and clothing to the humble mouse pad. It is even in the Liverpool FC and Kuala Lumpur City jerseys I like to wear. The material is resistant to water (it is made from petroleum, after all). Thus, it keeps the athletes' bodies dry or simply prevents stains.

Not long after the PR coalition of KEADILAN, DAP, and PAS came into power in Selangor and Penang in 2008, the two state governments placed restrictions on plastic bag usage. Initially, it was controversial. Penang prohibited the use of plastic bags on Saturdays starting 2009. In the following year, Selangor introduced a no plastic bag day on Saturdays, where 20 sen was charged for using plastic bags. In 2011, even the Federal

Government followed suit by making Saturdays a no-plastic-bag day while Penang made it an everyday affair.

Unfortunately, the numbers show that although the recycling industry is growing in Malaysia, it is not expanding fast enough to catch up with the amount of plastic that goes into the trash can. In 2022, in Kuala Lumpur and Putrajaya, 210,966 metric tonnes of plastic ended up in the landfill. Less than a fifth of plastic waste was recycled. In fact, this is a global problem. This is because recycling is energy intensive and requires tedious sorting and washing. It is much cheaper to simply make new plastic instead.

Also, there are different types of plastics. The drinking water bottles as well as polyester material are normally made from PET. Milk cartons and detergent bottles are made from HDPE. Straws, bottle caps, and medicine bottles tend to be made from PP. All three types of plastics can be recycled relatively easily. Jerseys for many of the top football clubs today, for example, are made entirely from recycled plastic bottles as they all come from PET. The 2023 All Blacks (New Zealand) rugby jersey is made from 93 per cent recycled plastics. Liverpool FC has been making jerseys entirely from recycled plastics since 2020.

But there are also other alternatives that must be seriously considered, not as a first resort, but as an option in the journey to reduce single-use plastics. One of them is truly biodegradable plastic. Many so-called biodegradable plastics are only compostable in highly controlled environments with exact temperature, pressure, and presence of nutrients, among others. So, unlike a piece of paper or organic waste, it does not biodegrade if left alone in the environment. Another type that is not truly biodegradable is oxo-degradable plastics that degrade with sunlight and oxygen, but still contain microplastics.

There is, however, truly biodegradable plastic. Malaysia's standards agency SIRIM has started an eco-labelling for biodegradable plastic with the international player Polymateria that manufactures it here in Malaysia. This technology allows—after a set time, say two years—for a plastic product to lose its physical properties and turn into a biodegradable wax that reacts with naturally occurring bacteria and fungi. A Malaysian company, Maribumi Starchtech, manufactures plant-based alternatives to plastics that are compostable within 180 days.

Highly Polluting Industries

One of the challenges we face in the journey towards net-zero is how to decarbonize highly polluting industries, often referred to as the hard-to-abate sectors: construction, steel, aviation, and others. Any reduction in the emissions from these industries will have a big impact.

Under the new government, high-voltage industrial consumers are allowed to install solar panels for self-consumption. Previously, this was not permitted, as these industries were major consumers of TNB, and it would have hit the utility's revenue. Many concrete and steel factories advocated for this change, particularly after we removed subsidies from electricity for major industrial consumers. This allows them to reduce their electricity bills while also fulfilling demands from their consumers to reduce their carbon emissions and make their businesses more sustainable.

In 1986, in the midst of a national economic crisis, Press Metal Aluminium was established in Puchong. The family that owns it had been in the business of hardware and trading before. Paul Koon, one of the family's seven children had just graduated from the US and was struggling to find for a job. As the family had been involved in aluminium trading, Koon felt that manufacturing aluminium extrusion—a material widely used in various industries—was something that he could participate in. He started Press Metal with only twelve workers.

The company grew rapidly but began to face stiff competition from China-based manufacturers. They decided that rather than competing with their rivals, they would start manufacturing operations in China in 2005. Two years later, the company entered the smelting industry by learning through the acquisition of a Chinese smelter. Today, Press Metal has five smelting and manufacturing plants in China, Selangor, and Sarawak. Most importantly, it takes up 70 per cent of the 2,400 megawatts produced by Bakun dam. Smelting is a very electricity-intensive industry, due to the electrolytic process required. Today, the company has one of the lowest costs of production of aluminium in the world due to the low tariffs for hydroelectricity. While the capital outlay of building a dam to produce hydroelectricity is big, operationally it does not require expensive fuel to run, only water.

As mentioned earlier, hydrogen also plays a key role for powering energy-intensive industries such as steel and cement. Today, steel and cement manufacturing heavily rely on coal. Currently, for every one tonne of cement, 500 kilogrammes of coal are required. Steel is traditionally made using blast furnaces, which reduces iron ore into iron. This process can be traced back to ancient China. It is powered by coke, a high-carbon material made by distilling coal. Blast furnaces consume a lot of space and are major carbon emitters.

Many contemporary Malaysian steel manufacturers use electric arc furnaces as a cleaner alternative. A heat-generating arc is created when electrodes are powered. The heat can be dialled up faster than blast furnaces. They are cheaper to set up but, as of now, are more expensive to run. Instead of using iron ore, they use recycled steel, scrap metal, or sponge iron. Globally, steel has a high recycling rate.

Global Boiling

For the longest time, I used to think that 'global warming' was something abstract. After all, Malaysians are accustomed to our hot and humid climate. While my family was comfortably middle-class, I slept with a small, boxed fan for much of my childhood. My father only bought an air conditioner for my room when I was fourteen years old. Hot and sweaty nights had been the norm until then. Similarly, during my first year at MCKK, we were only provided a small, rotating fan attached to the ceiling in our dormitories. The short break between classes, lunch, and prayer in the afternoon would normally see us taking off our shirts, allowing our sweat to dry, and catching a quick nap in the dorms. Our uniforms included a shirt and tie. The prefects (I, fortunately, was not one) had to wear a maroon jacket as well. Maybe we only survived because the temperatures, while still warm, were slightly cooler, back then.

In the last few years, however, it has been apparent that the world has become much warmer than usual. Six out of the ten most damaging wildfires in California took place between 2020 and 2021. In the latter year, massive wildfires took place throughout the world, from Siberia, Türkiye, Greece, Italy, North India, Western Canada, California, Algeria, and Tunisia. In 2021, wildfires in Arctic regions of North America and Eurasia emitted 1.76 billion tons of carbon dioxide, 150 per cent higher than the average annual carbon emissions between 2000 and 2020.[79]

When I began writing this book in 2023, I could truly feel the heatwave in Malaysia as temperatures surged to nearly 40 degrees Celsius. WMO announced that the globe was experiencing the hottest recorded temperatures in history. Normally, heat stress is measured in working environments when temperatures go above 32 degrees Celsius. July 2023 was the world's hottest month on record and, for four months in a row, the global ocean surface temperatures reached the highest in their records.[80] The UN Secretary-General Antonio Guterres declared that the era of

---

[79] Bo Zheng et al., Record-high CO_2 emissions from boreal fires in 2021. *Science* **379**, 912–917(2023). DOI:10.1126/science.ade0805

[80] WMO, 'July 2023 Confirmed as Hottest Month on Record', August 14, 2023, https://public.wmo.int/en/media/news/july-2023-confirmed-hottest-month-record [Accessed August 27, 2023].

'global boiling' had replaced global warming.[81] Just north of the border, Thailand was experiencing temperatures of over 40 degrees Celsius, reaching 45.4 degrees in the city of Tak, the highest ever recorded in the country's history.

In April 2023, Malaysians were shocked by the death of eleven-year-old Muhamad Syamil Aqil from Bachok, Kelantan, due to heatstroke and dehydration. The mother recalled how the doctor treating Syamil had informed her that the boy's kidneys and heart had been impacted by the heat. Syamil had been seen happily cycling around the kampung on the first and second day of Hari Raya Aidilfitri. On the third day, however, he developed a fever and experienced diarrhoea. The following day, the boy suffered from cramps and died at a clinic. This was followed by the death of one-year-old girl Nur Imani Ahmad Faris Fazli in Kota Bharu, Kelantan, from heatstroke. She began coughing and vomiting. Her organs were reportedly dehydrated and shrunk. After going to a clinic and then hospital, the doctor at the emergency ward only prescribed her medicine and asked the parents to bring her home. From 16 April to 10 June 2023, the Ministry of Health detected thirty-nine reported heat-related illnesses, including heat exhaustion, heat cramps, and heatstroke.[82]

Often, the elderly, unwell, and toddlers are more vulnerable to heatstroke due to their vulnerable conditions. But young ones are less experienced in dealing with heat, understanding the need to avoid the sun, and to drink plenty of water. As the examples show, they tend to be the first to lose their lives in a heatwave.

---

[81] UN News, 'Hottest July Ever Signals "Era of Global Boiling Has Arrived" Says UN Chief', July 27, 2023, https://news.un.org/en/story/2023/07/1139162 [Accessed August 8, 2023].

[82] 'Health Ministry Records 39 Heat-Related Cases, Says Dr Zaliha', *Malay Mail*, June 16, 2023, https://www.malaymail.com/news/malaysia/2023/06/16/health-ministry-records-39-heat-related-cases-says-dr-zaliha/74670 [Accessed September 4, 2023].

Annually, between June to August, El Niño results in unusually warmer and drier weather in maritime Southeast Asia, Australia, and parts of the Pacific. An El Niño cycle can last between two to seven years. The cool phase is La Niña. Research suggests that climate change and El Niño lead to even more extreme weather patterns. At the end of May 2023, I chaired the National Haze and Dry Weather Main Committee to prepare for the dry season and El Niño. Other than the technical departments under my ministry, the committee also involves other ministries: education, health as well as youth and sports. The education ministry orders schools to stop outdoor activities when they experience heatwaves. If the temperature exceeds 37 degrees Celsius for three consecutive days, schools are allowed to close.

On 11 May, partly due to surging temperatures resulting in an increase in demand for air conditioning, Peninsular Malaysia recorded its highest peak load of 19,716 megawatts (over 500 megawatts higher than the figure for 2022). But Peninsular Malaysia had ample reserve margins.

However, that same month, my social media was bombarded by complaints regarding increasing blackouts in parts of Sabah. A combination of factors contributed to this—a heatwave, a noticeable gap between decent reserve margins along the west coast of the state compared to an energy deficit along the east coast because most of the power generation happens in the former, a grid that relies solely on a single, 275 kilovolt connection between the two coasts, and the breakdown of several power plants. Faced with these challenges, the electricity operator, SESB had no choice but to ration electricity to prevent a more extended and widespread power disruption.

We also had to closely monitor water levels in dams providing water supply. SPAN opened a war room for this purpose. This is critical for states that do not have adequate water reserves like Kedah and Kelantan. NADMA, the Malaysian Armed Forces, and other agencies prepared for cloud seeding, should it be required.

Malaysia began to experience a weak El Niño in June, and the total amount of rain reduced by up to 40 per cent. Rice yields dropped following El Niño and the haze in 2023. In July, India banned the export of rice to reduce prices and promote food security for their local market. Malaysia was one of the countries affected by the ban. Out of almost 2,400 rice farmers based in Sungai Besar, Selangor, nearly 10 per cent complained of heat-related illnesses, which had not been common previously.[83] The Malaysian Palm Oil Board estimated that the heatwave in 2023 would reduce Malaysia's crude palm oil production by 3 million metric tonnes that year, a reduction of over 15 per cent since the previous year.

Handling the Haze

Peat is formed using partially decaying plant matter. It is the first phase in the formation of coal. Peat is used as fuel in some parts of the world although it produces heavy, sooty smoke that contributes to haze. At the same time, peatlands, whether those open for agriculture or those that are still preserved as forests, are a major risk when burned openly. The Malaysian government has invested in infrastructure in high-risk areas in Selangor, Pahang, Sarawak, Terengganu, and Johor. This includes digging tube wells to supply water to extinguish peat fires, building check dams to raise the water level to prevent fires in the first place, and monitoring via physical enforcement as well as using drones. As of June 2023, 102 tube wells have been built. A special Forest Fire Fighting Team has been established by the FDPM. We can see progress in the reduction of forest fire hotspots in Malaysia, since 714 hotspots were detected in Malaysia in 2023 compared

---

[83] Bernama, 'National Food Security Threats Likely as El Nino Returns', *Astro Awani*, October 12, 2023, https://www.astroawani.com/berita-malaysia/national-food-security-threats-likely-el-nino-returns-441430? [Accessed February 25, 2024].

to the two recent hot and dry seasons—2015 with 2,652 hotspots and 2019 (until October) with 2,091 hotspots.

Image 17: Going down to the site of peat fires in Johan Setia, Selangor. Peat fires are a major cause of haze.

In the middle of 2023, I attended a meeting of the ASEAN Sub-Regional Ministerial Steering Committee on Transboundary Haze Pollution. The sub-regional group consists of Malaysia, Thailand, Singapore, Brunei, and Indonesia, where the transboundary haze issue is largely interconnected. The countries have made much progress in managing haze and peat fires. There used to be a lot of finger-pointing in the past, but today, Malaysian companies operating in Indonesia have cooperated well with the authorities to reduce the problem of peat fires that lead to haze. However, I reminded everyone at the meeting that the alleviation of the haze issue in the region in the past three years may be attributed to the wetter and cooler La Niña period and the Covid-19 pandemic grinding much of the economy to a halt.

The first major transboundary haze incident in Southeast Asia took place in 1997, just as the region was facing the Asian

Financial Crisis. Indonesian farmers were still using traditional slash-and-burn methods to clear forests. But as logging reached an industrial scale, peatlands were drained out to prepare for palm oil plantations. Combined with the devastating El Niño conditions, this led to massive forest fires in Sumatra and Kalimantan. Other than countries in the ASEAN sub-regional group above, the haze reached the Philippines and Sri Lanka at its widest extent. Increasing healthcare expenses and disruptions in business activities, including flights, were estimated to have cost the region US$ 9 billion. A Garuda flight crashed and killed all 234 crew and passengers as a result of poor visibility.[84] Hospital cases related to air pollution increased by almost a third in Singapore.[85]

In 2015, the transboundary haze reached all the way to Vietnam, Cambodia, and the Philippines. In October that year, the PSI in Central Kalimantan reached a staggering 1,801. Under the measurement, 0–100 are healthy figures, while anything above 400 is deemed as hazardous. Schools were closed in Malaysia, Singapore, and Indonesia while the Kuala Lumpur Marathon had to be cancelled.

In 1997, among the measures taken by the Malaysian government was sending firefighters to Indonesia to assist in putting out the forest fires under Operation Haze. Our firefighters spent twenty-five days there. A similar operation, on a smaller scale, took place in 2005. METMalaysia has also set up the FDRS, modelled on a Canadian system, following the 1997 haze incident. The system monitors risk of forest fires and the data is shared with the ASEAN countries.

---

[84] Mydans, Seth, 'Indonesia Jet Crash Kills All 234 Aboard; Haze Was a Possible Cause', *The New York Times*, September 27, 1997, https://www.nytimes.com/1997/09/27/world/indonesia-jet-crash-kills-all-234-aboard-haze-was-a-possible-cause.html [Accessed March 6, 2024].

[85] Emmanuel, Shanta Christina, 'Impact to Lung Health of Haze from Forest Fires: The Singapore Experience', *Respirology (Carlton, Vic)*, vol. 5. pp. 175–82.

AATHP was signed in Kuala Lumpur in 2002. The ASEAN meteorological centre in Singapore monitors and assesses weather patterns and data from the region. A haze fund was established with contributions from member states to operationalize the agreement. To a certain extent, ASEAN departed from its traditional opposition to binding agreements on this matter. The agreement brings with it obligations for member states. The agreement is not perfect, as it highlights the often global challenges to overcome environmental and climate problems. These require international treaties, and the parties' commitment to the agreement is crucial for these treaties to succeed.

The limits of the agreement became clear in October 2023, as haze returned to the region. The data came from the ASEAN Specialised Meteorological Centre (ASMC) in Singapore. When we revealed that the air quality was deteriorating in spots across Malaysia, we also listed down the hotspots across the region—in Sumatera and Kalimantan, Indonesia. That is why the worst haze was along the west coast of Peninsula Malaysia and Sarawak, which are located near the hotspots. Indonesia's environment and forestry minister, Siti Nurbaya Bakar, denied that our haze had anything to do with her country: 'The fact is that there is no transboundary haze [. . .] They [Malaysia] refer to hotspot data? Don't they know the difference between hotspots and fire spots? If [you] don't know exactly, don't talk carelessly.'[86]

This is the age of the Internet. There was a time when even the air pollution index was covered by the Official Secrets Act in Malaysia. But, as more sources of information—including on air pollution and satellite data of global hotspots—are available

---

[86] AFP, 'Malaysia Blames Indonesian Fires for Haze, Poor Air Quality', *The Jakarta Post*, October 1, 2023, https://www.thejakartapost.com/world/2023/09/30/malaysia-blames-indonesian-fires-for-haze-poor-air-quality.html [Accessed March 6, 2024].

online, the public can draw their own conclusions on what was happening. This was not the first haze-related incident. The preceding ministers of environment, Wan Junaidi Tuanku Jaafar and Yeo Bee Yin, had to handle two bad haze incidences in 2015 and 2019, respectively.

Obviously, ordinary citizens across the region, not just in Malaysia but also in Singapore and Indonesia, are tired of the haze. It cannot be normalized. Therefore, the existing ASEAN agreement and forums must be updated to meet the current needs and expectations of ordinary citizens on this matter.

During the 2023 haze incident in Malaysia, as of October 7, asthma cases increased two-and-a-half times compared to pre-haze numbers. Cases of upper-respiratory tract infections increased two times. And the number of patients for conjunctivitis doubled.

I wrote a letter to Siti Nurbaya, offering Malaysia's assistance to Indonesia to fight the forest fires blazing there. There have been many calls for Malaysia to emulate Singapore, which introduced a Transboundary Haze Pollution Act in 2014. The legislation allows the Singaporean government to act against any companies responsible for causing haze in the island republic. However, from the information I gathered, while the Singaporean government has tried to act against a few companies, successful prosecution remains a challenge. There are issues of collecting evidence and sharing information—what can Malaysia do if authorities in Indonesia do not cooperate?

We have the example of the so-called Acid Rain Agreement between Canada and the US. The Canada–US Air Quality Agreement was signed in 1991. The transboundary treaty initially focused on addressing acid rain but eventually expanded to include smog. This was based on the willingness of the two countries to enter into the agreement. A bilateral International joint commission had been handling issues regarding common water bodies between the two countries along with an air quality committee. To make the AATHP effective, the Acid Rain Agreement can be a working example.

On the other hand, it must be noted that while burning methods cost farmers in Indonesia US $180 per hectare, non-burning methods cost more than four times that amount. This shows that there will always be a limit to how ASEAN or specific neighbouring countries, like Singapore and Malaysia, can contribute to fight the haze in Indonesia. At the same time, the massive cost difference makes it tempting for smallholders there to opt for the dirtier method. Maybe what Singapore and Malaysia can consider is to pay for a fund that encourages smallholders to move away from burning the stubble to clear the farmland.

Climate Adaptation

Often, while discussing environmental protection and climate resilience, mitigation—such as reducing carbon emissions and increasing forest cover or reducing deforestation—is the focus. Not enough attention is given to adaptation.

We definitely need to play our part in mitigating our emissions, but Malaysia's carbon footprint is very small in the grand scheme of things. Adaptation, on the other hand, is crucial to deal with the irreversible damage to nature that we have caused. This requires predicting the impact of climate change and making the necessary adjustments. There is a wide scope of choices: barriers against sea level rises; early warning systems for landslides, floods, and tsunamis; increasing tree cover and green spaces in cities; insurance schemes that reflect changing climate needs; as well as improving methods to deal with food waste. As mentioned in the previous chapter, there needs to be grey and green infrastructure to deal with climate change.

In 2023, the Cabinet approved the public sharing of flood hazard maps. This allows potential homebuyers to identify areas with high flood risks, aiding them in making informed property purchases. There has been a lot of resistance from developers, who have previously profited from the lack of transparency by building on flood plains. Banks and insurers can also use the data to calculate risk accordingly.

It is estimated that by the end of the century, parts of Malaysia will see 23.6 per cent more rainfall. It will also be more intense. This means that much of the grey infrastructure that has been built to prevent floods will not be sufficient to deal with the changing weather patterns.

In January 2023, I launched PNBCAP. PNBCAP looks at the problems of heat islands and flooding faced by the island. Among the measures that have been planned are increasing the number of tree-lined streets, creating pocket parks to develop microclimates and reduce hard surfaces, greening car parks, and urban farming. Blue-green corridors have also been created to allow for greater retention of storm water before it goes into rivers as well as the creation of upriver retention ponds. PNBCAP also collaborates with the Ministry of Education for youth and students' programmes with a strong emphasis on women and girls through the Penang Climate Board and civil society.

Under the Kyoto Protocol, PNBCAP received a grant of US $10 million as part of the Adaptation Fund. The project also led to Penang winning the Climathon Global Cities Award in 2020, after being shortlisted alongside the likes of Miami (US), Salvador (Brazil), and Dublin (Ireland). PNBCAP has become an example of how future collaborations with other cities regarding climate adaptation can be done.

Malaysia, through the NRES, has been working on a national adaptation plan using grants from the Green Climate Fund. It aims look at public health, agriculture, and food security, forestry and biodiversity, water resources and security, as well as infrastructure and cities. It deals with the urban heat island effect and incorporates early warning systems for disasters. It also looks at combining flood and slope hazard maps to ensure all developments incorporate the necessary risks in planning. The plan seeks to include nature-based solutions as alternatives to grey infrastructures where possible. This is in-line with Prime Minister Anwar Ibrahim's focus on climate resilience as a key plank of his economic vision.

Forests and Wildlife Conservation

> The woods on Ox Mountain were once beautiful. Because they were on the edge of a large country, they have been attacked with axes and hatchets, so how could they remain beautiful? [. . .] People seeing its denuded state assume that it never had been otherwise, endowed with rich resources. Yet how can this state be the true nature of this mountain?
>
> —Mencius, ancient Chinese philosopher

One of the first books of the nineteenth century to express concern over the impact of humans on the environment was Alfred Russel Wallace's *Tropical Nature and Other Essays* published in 1878. Today, it is seen as the precursor to Rachel Carson's *Silent Spring*, which was published almost a century later. Unlike the latter, which brought about the contemporary environmental movement, not much popular attention was given to Wallace's book.[87]

Wallace was better known for his other work, *The Malay Archipelago*. In it, he records his explorations, beginning from the southern Malay Peninsula. He started in Singapore, going north to Malacca and Johor. He eventually travelled all the way to Papua New Guinea including stops in Sarawak, Java, and the Celebes. He covered a distance of over 22,000 kilometres and collected

---

[87] Wallace, Alfred Russel, *Tropical Nature and Other Essays*. London: Macmillan & Co, 1878.

125,000 plant and animal specimens. In the year before he left for his expedition to the East, Wallace met Charles Darwin. In 1855, Wallace came up with the Sarawak Law: 'Every species has come into existence coincident both in space and time with a pre-existing closely allied species.'[88]

Three years later, Wallace and Darwin jointly presented a paper on evolution by natural selection, a year before Darwin's famous *The Origin of Species* was published. In his expedition to the Malay Archipelago, Wallace was assisted by a Malay man from Sarawak, Ali. While Ali started out as a cook and teacher of Malay for Wallace, eventually he played a key role in collecting 5,000 birds and other specimens, making a major contribution to *The Malay Archipelago*.[89]

The notorious mafia boss Al Capone, who terrorized the US in the Roaring Twenties, was nicknamed 'Scarface' after he was slashed on the left side of his face. The nickname became the title of young author Armitage Trail's novel in 1930. Two years later, the movie adaptation came out. It was remade in 1983 starring Al Pacino and Michelle Pfeiffer. Interestingly, both movies are now considered among the top gangster movies of all time.

Early one morning in the 2000s, Mek Jah Ismail, a woman in her mid-sixties from rural Jeli, Kelantan, was tapping rubber in her smallholding. A Malayan tiger surprised the lady and attacked her. Despite falling repeatedly while trying to escape, she managed to slash the tiger's face a few times using her rubber-tapping knife. She was badly wounded on her head, neck, and chest. The tiger managed to run away after being wounded. A few months later,

---

[88] Wallace, Alfred Russel, 'On the Law Which Has Regulated the Introduction of New Species (1855)', *Alfred Russel Wallace Classic Writings*, Paper 2. 2009, https://digitalcommons.wku.edu/dlps_fac_arw/2 [Accessed February 14, 2024].

[89] van Wyhe, John and Drawhorn, Gerrell M., '"I am Ali Wallace": The Malay Assistant of Alfred Russel Wallace', *Journal of the Malaysian Branch of the Royal Asiatic Society*, vol. 88, 1. 2015, pp. 3–31, https://www.jstor.org/stable/26527691 [Accessed January 21, 2024].

another rubber tapper was found with his head severed from the body after being attacked by the same tiger. Another victim was found with his head and hand bloodily separated from the body. Due to Mek Jah's act of self-defence, the tiger was named Scarface after Al Capone. He killed a total of four people and only Mek Jah survived. If Hollywood is interested, a movie portrayal of Kelantan's Scarface would have all the ingredients of a blockbuster like *Jaws* or *Silence of the Lambs*.

Scarface was sent to the National Wildlife Recovery Centre, a temporary settlement for seized wildlife in Sungkai, Perak. Many of the tigers there have been caught by the wildlife department due to conflicts with humans. Some are rescued from the illegal wildlife trade. Maneaters are not allowed back into the wild. But their offspring can be reintroduced tigers into the wild. Scarface bred with another tiger, Tanjung, and produced an offspring, Sungkai—named after his birthplace.

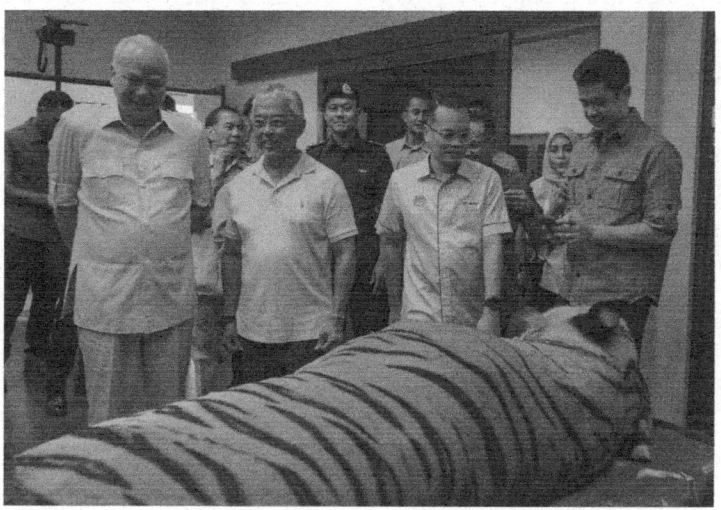

Image 18: From left to right: Sultan Nazrin Shah of Perak; The XVIth King of Malaysia, Al Sultan Abdullah Shah of Pahang; and Regent of Pahang, Tengku Hassanal Shah at the National Wildlife Rescue Centre in Perak. We were briefed on tiger conservation techniques.

Some of the offspring of the Sungkai tigers have been transferred to the National Tiger Conservation Centre in Lanchang, Pahang. The plan is for this to be the final facility before tigers bred in captivity are released to the wild. This is a delicate process that needs to be done properly. Previously, releasing captive animals back to the wild was not seriously considered, but today, it is seen as one of the possible options in our desperate attempt to preserve endangered animals in their natural habitat.

In July 2023, I joined the then king, Sultan Abdullah, and the regent of Pahang, Tengku Hassanal Ibrahim Alam Shah, for a historic moment: the declaration of the Al Sultan Abdullah Royal Tiger Reserve. Sprawling over 134,000 hectares, it is about a quarter of Bali's size and just slightly smaller than Greater London. As part of the event, we visited the Lanchang facility. First, we were shown how tigers were fed fresh meat (beef or pork) in their cages. Tigers in the cages next to them were allowed to chase live chickens. Then, we were taken out to a large, fenced area. The piglet of a wild boar had been released into the area. Here, another tiger that is more independent was supposed to be practising hunting its prey. This was a trial run for life in the wild. With a royal audience, we saw the piglet run in circles in the fenced area chased by the excited tiger. The piglet squealed at a high pitch. Then, just as the tiger was about to catch it, the piglet turned around and squealed loudly. The tiger immediately turned around and ran away from the piglet! This tiger clearly needed a lot more practice if it was to not go hungry in the forest. We were able to see another tiger successfully chase and devour a deer.

To the north of the state of Perak, bordering Thailand, lies the Royal Belum state park, another major habitat for Malayan tigers today. This massive forest reserve is one of the oldest rainforests in the world, estimated to be 130 million years old. This means it has been around in the age of the dinosaurs and is older than the Amazon and Congo rainforests. Spread across approximately 300,000 hectares, the Belum–Temenggor forest complex (of which Royal Belum is a substantial part) is more than four times the size of Singapore. Other than the tigers, the forest is inhabited by a rich array of large animals, such as the Asiatic elephant, Malayan gaur, Malayan tapir, and Indochinese leopard. It is the home to more than eighty species of hornbills. Rare tropical hardwoods, such as *meranti* (shorea), *chengal*, and our national tree, *merbau* (Malacca Teak) flourish in Royal Belum.

In 1909, the jurisdiction of the area was transferred from the Reman Kingdom, a Malay tributary state of Siam, to Perak, which was administered by the British Federated Malay States. The British Resident for Perak Ernest Woodford Birch made a sixteen-day trip into Belum-Temenggor. There were people residing in Kampung Belum there, and Birch wanted to inform the community that the area was no longer protected by the Kingdom of Siam but was part of Perak (and thus, a British protectorate). It took him and his entourage nine days to reach the village by foot and on elephants. The community relied heavily on elephants for transportation and to assist them in clearing land to build their kampung.

After the Second World War, in 1952, at the height of the Communist insurgency, the residents of Kampung Belum Lama were instructed to move to safer territory. Over 200 villagers walked barefoot for over 80 kilometres in a week's time. They crossed into Thailand, and then into Kelantan. Today, the journey is easier, as a lot of the forests have sunk beneath Lake Temenggor and thus, can be crossed in a boat instead of on foot. I was told that if one journeys to the ruins of the village, the remnants of elephant baths and graveyards can still be seen.

In Sabah, the Danum Valley is estimated to be as old as Royal Belum. Its area is 44,000 hectares, and it's situated at the north-eastern part of Borneo. Considered a complex ecosystem, this valley is a treasure trove of biodiversity. Meranti trees reach up to over 100 metres in height, among the tallest in the tropics. Just in 2016, around thirty trees over 90 metres tall were detected there. Further inland is the larger Maliau Basin, a huge bowl-shaped river basin that has remained virtually a pristine jungle. Gibbons, orangutans, Bornean banteng, Malayan sun bear, and Bornean pygmy elephants roam these areas.

Image 19: Visiting Sepilok Orangutan Rehabilitation Centre in Sabah, which rehabilitates orangutans caught in human–wildlife conflicts due to deforestation as well those being kept as pets. The orangutans will then be released back into the wild.

Malaysian forests are dominated by dipterocarp trees. The name refers to the fruit, which has prominent 'wings'. The fruits only appear once in several years. When they fall from the trees, they can stay in the air longer due to their unique shape, which allows them to move further to propagate. A family of hardwood trees, they provide a canopy that allows other trees to grow. Sabah, Sarawak, and Brunei have the greatest variety in dipterocarp species, which have dominated their jungles for a million years.

Malaysia is blessed to be one of the most megadiverse countries in the world. A fifth of the world's animal species are estimated to be found in the country. Other than being home to diverse species, another requirement to be defined as a megadiverse country is possessing endemic species—species that can be found only in a defined geographic location. The coasts of Sabah are part of the Coral Triangle, located between the Pacific and Indian Oceans, stretching between Luzon Island in the Philippines to the north, Bali Island in Indonesia to the southwest, and the Solomon Islands to the south-east. It is less than 2 per cent of the world's oceanic area, yet, it contains 76 per cent of the known coral species.

Hydroelectricity in itself does not rely on fossil fuels or produce carbon emissions. But building large dams in a country like Malaysia where almost all the riverheads in the highlands are covered with tropical rainforests means the destruction of trees and displacement of wildlife. It also displaces indigenous communities who live, hunt, and collect forest products in the area. In the 1970s and 1980s there was a proposal to build a hydroelectric dam inside our national park, Taman Negara. It provoked a massive campaign by the burgeoning environmental movement in Malaysia that finally led to the Cabinet reversing the decision.

When I was NRECC minister, I inherited oversight over some large hydroelectricity projects that were approved prior to my time and that were already under construction. Moving forward, however, I stated that we should look at alternatives for renewable energy other than large, hydroelectricity dams. We can still generate power from small hydroelectricity plants using the run-of-river method. This does not involve submerging enormous areas of forests underwater.

1992 Earth Summit

During the 1992 UN Earth Summit at Rio de Janeiro, Malaysia committed to maintain 50 per cent forest cover. The Convention on Biological Diversity and the UNFCCC were signed at the conference. This was a recognition that while humankind has been using more fossil fuels since the Industrial Revolution, filling our atmosphere with more carbon emissions; we have also destroyed large areas of our forests, which absorb the carbon.

Oxford economist and environment specialist Professor Dieter Helm wrote pointedly:

> While pumping out ever more carbon, we have simultaneously been reducing the ability of the natural environment to absorb it. The capacity of nature to mop it up by natural sequestration through absorption by trees, soils and peats has been decimated over the past 30 years.[90]

At that same event, Malaysia was vocal in its opposition to what then Prime Minister Dr Mahathir Mohammad termed as 'eco-imperialism'. Malaysia took the lead on the matter. While one can see this as a tactical posturing by Mahathir on one hand, the underlying principle that the developed world should be more committed and assist the developing world in conserving our forests while pursuing our own development objectives is an enduring principle (more of this in the following chapter).[91]

However, an added challenge for Malaysia's forest conservation is that land and forests, just like water, fall under the State List. State governments guard these rights jealously. Unlike states in the US or Australia, they do not have police forces or schools under their jurisdiction. Along with Islam, land, forests, and water are the few substantive rights within the jurisdiction of

---

[90] Helm, Dieter, *Net-zero: How We Stop Causing Climate Change*. London: William Collins, 2021, p. 2.

[91] Keeble, Brian, 'Reflections on the Earth Sumit', *Medicine and War*, vol. 9, 1. January–March 1993, pp. 18–23, https://www.jstor.org/stable/pdf/45354685.pdf [Accessed January 21, 2024].

state governments. National parks and wildlife fall under the list of powers shared by the Federal and state governments.

Malaysia is one of the world's largest producers of tropical timber. The British were not as interested in timber compared to mining or rubber planting. As a result, the laws were not unified, and gaps existed between the different sets of laws. In 1972, the National Forestry Council was formed to bring different state governments to the table to sit with the Federal Government. Five years later, the National Forestry Policy was launched, where reforestation was promoted.

The various state laws were brought together and were eventually passed in 1984 as the National Forestry Act that regulates the management, conservation, and use of forests. Permanent forest reserves were defined and categorized for the first time.

As of 2022, our forest cover stood at 54 per cent. A year later, amid sobering news that tropical forest loss was accelerating globally, Indonesia and Malaysia managed record low levels of deforestation.[92] After peaking between 2009 and 2016, primary forest loss in Malaysia has been low for the past few years, and according to the study, both government and corporate initiatives contributed to this.

Mangroves are an even superior carbon sink than forests on land, thereby playing an important role in carbon mitigation. They grow on the coast in saltwater and brackish water. But mangroves are also a climate adaptation asset: they protect us from coastal erosion and storms. They provide a habitat for juvenile fishes and shelter corals from heatwaves. After Indonesia and Brazil, Malaysia has the third largest mangrove forest in the world.

---

[92] Weisse, Mikaela, Liz Goldman, and Sarah Carter, 'Tropical Primary Forest Loss Worsened in 2022, Despite International Commitments to End Deforestation', *Global Forest Watch*, June 27, 2023, https://www.globalforestwatch.org/blog/data-and-research/global-tree-cover-loss-data-2022/ [Accessed July 22, 2023].

It is no surprise that Southeast Asia has the most diverse type of mangroves known to man.

For many in Southeast Asia, Boxing Day 2004 is a date that we cannot forget. A major earthquake measuring 9.1 on the Richter scale hit the north-west coast of Sumatra. It created a massive tsunami of up to 30 metres affecting Aceh in Sumatra, Sri Lanka, India, and Thailand. It was one of the worst disasters of modern history, killing around 230,000 people. Malaysia was not spared: the Boxing Day tsunami reached Penang, Kedah (including Langkawi), and Perak. The long-term environmental impact was also significant: damaged mangroves, coral reefs, and pollution from damaged infrastructure. Saltwater got into rivers, lakes, and groundwater. Scientists, however, found out that communities shielded by mangroves survived the disaster better than those where they had been cleared.[93]

In 2023, the BBC interviewed an elderly Indonesian man, Nurhadi, involved in harvesting mangrove wood to turn it into charcoal. He said that there was not much money in the business, but it was merely a means for half of the people of Batu Ampar on Kalimantan to survive and to put food on the table. The impact of such deforestation is clearly felt. At the turn of the century, there were only ninety charcoal furnaces, in 2023 there were 400 more.[94] People have ventured into protected areas, beyond the legal ones, to harvest mangroves. Charcoal is used in metallurgy, filtration, construction, and even healthcare. And, of course, to be made into briquettes for barbecue! Due to its value as a metallurgical fuel in the production of iron, it drove

---

[93] Zimmer, Katarina, 'Many Mangrove Restorations Fail. Is there a Better Way?' *Knowable Magazine*, July 22, 2021, https://knowablemagazine.org/article/food-environment/2021/many-mangrove-restorations-fail [Accessed August 22, 2023].

[94] Lumbanrau, Raja and Lorna Hankin, 'Burning Mangrove Trees for a Living: "I'd Quit Tomorrow if I Could"', *BBC*, August 15, 2023, https://www.bbc.com/news/world-asia-66393515 [Accessed August 21, 2023].

rapid deforestation in sixteenth century Europe, even prior to the Industrial Revolution.

The Significance of Wildlife

The Malaysian love affair with the durian is well known. Known for its pungent smell and creamy taste, the fruit is definitely an acquired taste. Anthony Burgess, the author of *A Clockwork Orange* who taught in Malaya, colourfully and imaginatively described eating durian as 'eating sweet raspberry blancmange in the lavatory'.

What is less well-known is that durians are almost exclusively and effectively pollinated by wild bats, not insects. Durian flowers bloom only once a year. Since this happens at night, they are best pollinated by bats, which are nocturnal. The mammal also hovers the longest over the flowers and does the least damage to the flowers. They put their long snouts into the flowers to get the nectar. Regardless of whether it gets pollinated or not, the flowers then fall to the ground.

But protecting bats is not only important for durians. When the Covid-19 pandemic broke out, it was part of an increasing trend of pandemics that had been emerging in recent years. Obviously, Covid-19 was by far the most devastating one. A study jointly conducted by the WHO and the People's Republic of China pointed to wild bats, which infected wildlife with the disease first and then humans. The reason pandemics are more common and why these viruses infect human beings is largely the destruction of wildlife habitat, lack of nutrition, as well as infections that cause bats to be stressed due to which they transmit viruses.[95]

In 1999, the Nipah virus killed over 100 people and led to the culling of a million pigs in Malaysia. The fatality of the virus is as

---

[95] 'WHO-Convened Global Study of Origins of SARS-CoV-2: China Part', WHO, 2021, https://www.who.int/publications/i/item/who-convened-global-study-of-origins-of-sars-cov-2-china-part [Accessed August 12, 2023].

high as 75 per cent. While we have not experienced any further cases, Bangladesh and India have experienced outbreaks on an almost annual basis. It has been suggested that this was due to El Niño and deforestation. The 1997–1998 El Niño phenomenon caused severe forest fires in Kalimantan and Sumatra, which then led to severe haze over Peninsular Malaysia. This may have affected the flowering and fruiting of forest trees. Combined with increasing deforestation, these factors possibly led to the migration of fruit bats (also known as flying foxes), which are the carriers of the Nipah virus, to fruit orchards. The earliest cases involved piggeries that were situated in fruit orchards.[96]

Iman, a female Sumatran rhino, died in 2019 at the Borneo Rhino Sanctuary in Lahad Datu, Sabah. It was the fifth rhino to die in five years in Malaysia. Sadly, it was also the last Sumatran rhino in the country. Thus, the species is extinct in the country. The habitat of Sumatran rhinos once spanned from India to China, but they are now critically endangered, left only in Sumatra and Kalimantan. It was only in 2006 that the first photo of the wild and elusive Sumatran rhino was captured on camera-traps, and the first video footage was recorded the following year. Conservation groups then convinced the local authorities to turn the area into a conservation zone but it was too late. The species is known for its reproductive challenges. Rhinos are extremely solitary animals for whom even mating is a challenge. They also have a long gestation period. But what has made the situation worse, of course, is the fragmentation of their habitats due to development.

Rantau Abang in Terengganu used to be well-known as a nesting site for leatherback turtles. It is the largest non-crocodilian

---

[96] Chua Kaw Bing, Chua Beng Hui, and Wang Chew Wen, 'Anthropogenic Deforestation, El Niño and the Emergence of Nipah Virus in Malaysia', *Malaysian Journal of Pathology*, vol. 24, 1. June 2002, https://www.mjpath.org.my/past_issue/MJP2002.1/anthropogenic-deforestation.pdf [Accessed January 22, 2024].

reptile and is found in oceans across the world. As a child, when I would go back to my parents' hometown in Kota Bharu, we would have boiled turtle eggs as a delicacy. They were slimy and tasted a bit of the sea. But now, this is illegal because the number of leatherback turtles has plummeted. They have few natural predators once they are fully-grown, but they are at risk in their early stages. Hatchlings are attracted to lights. Thus, man-made lighting disorients hatchlings who are trying to make their way to the sea. Leatherback turtles get stranded due to injury. The ingestion of balloons and plastic bags—which look like their food, jellyfish—are often fatal. The last known landing of these turtles happened in Rantau Abang in 2017.

Growing up in the 1980s and 1990s, these moments were poignant for me: I had been taught in school about rhinos in our forests and leatherback turtles on our beaches. Today, they are gone. The only distant (yet controversial) solace is science: the Borneo Rhino Sanctuary harvested Iman's egg cells with a plan to one day create an embryo. What has changed for Ilhan, my son? Which wild animals that he grew up knowing will disappear in his lifetime?

As more forests are cleared, the communities at the plantations and orchards developed in their stead find wildlife incursions, even fatal conflicts, becoming common. In Kluang in the southern state of Johor, Asian elephants are known to have entered the town, even coming up to people's homes. We are not talking about a frontier rural outpost but an urbanized town with a population of almost 300,000! One of the methods to protect plantations and orchards are electric fences, which fragments the space for wildlife. The Malayan tiger roams areas of up to 300 kilometres. In Sabah, the Bornean pygmy elephants can travel up to 50 kilometres in just a day.

Clearly, our hubris in thinking that we could tame nature in our thirst for profit has upset the balance of nature, and now nature is fighting back. Climate change aside, pandemics and human–wildlife conflicts are among the examples of nature's response to mankind for going beyond the planetary boundaries.

Image 20: The National Elephant Conservation Centre in Kuala Gandah, Pahang, provides an opportunity for the public to engage closely with wild and tame elephants.

While focusing on forest destruction and poaching as contributors to the survival of endangered species, it is important to know that fragmentation of habitats also plays a big role in exacerbating this issue. In Peninsular Malaysia, the authorities have recognized the importance of preserving the Central Forest Spine and restoring ecological corridors within the area. The construction of the Kuala Krai–Kuala Pilah Highway, also known as the Central Spine Road, involved the construction of the Sungai Yu Eco Viaduct to allow for wildlife to cross safely from Taman Negara and the Titiwangsa mountain range. The viaduct spans over one kilometre and cost over RM60 million. On the Second East West

Highway connecting Simpang Pulai in Perak to Kuala Berang in Terengganu, two wildlife crossings have been built. These crossings were planned accounting for elephant trails. Elephants break branches and feed on the plants along the trails and, as new shoots grow, they attract smaller mammals in their wake. Salt licks have been placed along the route as another attraction for the animals to use it.

On International Tiger Day 2023, I visited Royal Belum. I took the forty-minute boat ride across the man-made Temenggor Lake to the Sungai Kejar basecamp. I then trekked into the forest, passing lots specially planted grass for gaur. I was shown snares—made with simple wires bought from your neighbourhood hardware store—that were used to trap various wildlife. Camera traps there had recorded numerous three-legged tigers and sun bears. They lost their legs while trying to escape the snares. Some are not so lucky and die of infection and blood loss. I was then shown a salt lick. Just then, one of the Orang Asli rangers spotted fresh deer blood on the ground. We moved further inland and saw clumps of deer hair strewn across the ground. One of the rangers explained that this meant a deer had been dragged by a tiger. Thrilled, I jumped at the opportunity to look at tiger footprints when one of the sniper-wielding rangers stopped us because he suspected the tiger was still nearby!

I was excited to see that there were positive signs after all. In 2020, the number of Malayan tigers in the wild was estimated to be less than 150. There were an estimated 20 times more tigers in the 1950s. Videos of tiger cubs with their mothers, recorded by the camera traps, hinted at us still have a chance to protect the animal that symbolizes not only our football team but also adorns our national crest! After all, Royal Belum is the first location in Southeast Asia to be certified under CATS. This means that the reserve has reached international conservation standards for managing the wild tiger population. We are looking at getting other reserves the same certification. The government has, likewise, outlined several extraordinary actions to preserve our iconic apex predator.[97] The FAM and Malaysia's biggest bank, Maybank—both with tigers in their logos—contribute to the conservation of the Malayan tigers.

I also attended a dialogue with representatives of Orang Asli communities from all over Malaysia. They complained about the real issues they face living next to forests and wildlife. Why was the wildlife department complaining about lacking a budget to deal with human–wildlife conflict? The short answer, I told them, was that now, the conflicts have increased in frequency, and the financial demands on the department have gone far beyond the usual budget allocated to it by the Ministry of Finance. But, they then asked, why was it that the department kept releasing captured wildlife into territories where the Orang Asli lived and earned a living, exposing them to the risk of damaging their crops, injury, and worse, death? I did not have a straightforward and simple answer for them. These were heart-wrenching pleas of a community that used to live in harmony with nature but has been facing the havoc wreaked by the industrialized modern society and its extractive and exploitative economy.

In October 2023, a twenty-five-year-old Orang Asli named Pisie Amud living in Kampung Sugi, Gua Musang went out with his blowpipe to hunt. But he did not return home. Instead, he was discovered lifeless with violent scratches on his head and body while his left leg was missing. When two villagers found Pisie's body, they alerted the rest of the community and, after dinner, they took turns to guard the remains while waiting for it

---

[97] These include increasing boots on the ground for enforcement and monitoring, strengthening the conservation of tiger habitats, establishing the Wildlife Crime Bureau under the Royal Malaysian Police, enhancing the wildlife forensic laboratory and the accreditation standards for management of habitat, along with financial incentives for state governments.

to be retrieved. At around 4.00 am, the tiger returned, perhaps intending to salvage what remained of Pisie's body, but the villagers' made noises and lit firecrackers to scare it off. When he was buried, the community placed sharpened bamboos, which they believe would protect the grave from the tiger. When I went to Gua Musang the following month, I met Pisie's parents and his wife and children. I saw the deep sadness in their eyes. To see a young man's life taken away—coming face-to-face with a victims of human–wildlife conflict—left a deep mark on me. Stories like this reinforce the need to properly conserve wildlife—if that happens, tragedies like this will cease. Tigers do not normally consume humans, they only do so when they struggle to find food or are ill and feel threatened.

Failure is not inevitable. We can turn the tide. Look at India, which successfully brought the Bengal tiger back from the brink of extinction. The number of Bengal tigers in India was 40,000 in 1947, when it gained independence. In the 1970s, the number dwindled to a mere 1,800. The Indian government then introduced Project Tiger, aimed at protecting habitats and preventing poaching. At that time, there were nine tiger reserves spread across 18,278 square kilometres. In 2018, the number of tigers in India is estimated to be between 2,600 to 3,000. Today, there are fifty-three tiger reserves amounting to over 75,000 square kilometres. While the effort is not without its weaknesses, it has largely succeeded. The fact that India is home to 75 per cent of the world's wild tigers today is testament to that. It has also stemmed deforestation. In 2023, fifty years after launching Project Tiger, India initiated the International Big Cat Alliance to preserve the world's big cats: tigers, lions, snow leopards, leopards, jaguars, pumas, and cheetahs. I made it a point for Malaysia to join the alliance to share the best practices with India and other member countries.

Overall, forty-eight species have been rescued from extinction—a list that includes the Iberian lynx, Przewalski's horse, and the Puerto Rican Amazon, also known as the Puerto Rican parrot.[98] The Iberian lynx, a wild cat species from southern Spain suffered due to habitat loss, road accidents, and hunting. There were less than 100 lynxes in the wild in 2002. Following a series of rewilding programmes, the number shot up to 309 within nine years. Przewalski's horse, known as *takhi* among Mongolians, is the world's last wild horse species. These horses became extinct in the wild in the 1960s due to the disappearance of their habitats, inbreeding among horses in captivity due to a genetic bottleneck of a limited gene pool, as well as hunting. However, following successful captive breeding programmes, these horses were released back in the wild and their status was upgraded from 'extinct in the wild' to 'critically endangered' in 2005, and then to 'endangered' six years later. The Puerto Rican Amazon used to be widespread on the island but suffered from loss of habitat. This became worse in 1989 when Hurricane Hugo destroyed half of the population. After the establishment of a population of captive parrots, they were then reintroduced to forests. In 2017, Hurricanes Irma and Maria devastated the early rewilding efforts, but numbers remained strong in one of the habitats, and wild breeding continues to be recorded.

Earlier, I wrote how bats more effectively pollinate durians than insects. But insects pollinate hundreds of thousands of flowering plants. Overall, with regards to protecting biodiversity and the environment, insects play a crucial role in the ecosystem. Insects and bacteria process dead organisms into the soil. Malaysia is among the countries using the larvae of black soldier flies to process organic waste. After the Russian invasion of Ukraine led

---

[98] Bolam, Friederike C, Louise Mair, Marco Angelico, et al., 'How Many Bird and Mammal Extinctions Has Recent Conservation Action Prevented?', *Conservation Letters*, vol. 14, 1. September 9, 2020, https://conbio.onlinelibrary.wiley.com/doi/10.1111/conl.12762 [Accessed March 7, 2024].

to a global price hike for grains, affecting corn used as chicken feed in Malaysia, the government had been exploring the use of the same species of fly to feed chickens.

Termites create holes inside the ground, making it more permeable for water and easier for grass to grow. Being at the bottom of the food chain as a source of food for many animals, insects play a crucial role not only in the animal kingdom but also in the supply of food to humans. Some insects are pests but can be naturally controlled by *other* insects. Maggots are used to treat wounds, as they consume dead flesh. And let's not forget silk, honey, and other important products created by insects.

Insects have been around for over 300 million years. They are believed to have evolved from crustaceans. Insects have been hugely successful as a species. Weighed together, they are estimated to be seventeen times heavier than the weight of all humans. It was only since the Industrial Revolution that they faced a major challenge to their existence: humans. Today, insects are estimated to suffer a rate of extinction higher than mammals, birds, or reptiles. This is largely due to widespread agriculture and the use of pesticides, habitat loss, the growth of cities, and climate change.

In Puerto Rico, almost 98 per cent of ground insects are extinct. It is estimated that tropical insects are more vulnerable to the effects of climate change because they are used to a stable environment. As a result, the number of frogs has fallen by at least 50 per cent. The population of Puerto Rican tody, a brightly coloured endemic bird that almost entirely depends on insects, plummeted by 90 per cent.[99] Imagine the impact on predators that consume frogs and birds in the Puerto Rican forests.

---

[99] Carrington, Damian, 'Insect Collapse: "We Are Destroying Our Life Support Systems"', *The Guardian*, January 15, 2019, https://www.theguardian.com/environment/2019/jan/15/insect-collapse-we-are-destroying-our-life-support-systems [Accessed December 28, 2023].

In 2022, the UN Biodiversity Conference achieved a significant milestone, to the point that it was hailed as the equivalent of what the 2015 Paris Agreement was for climate change. This was the Kunming–Montreal Global Biodiversity Framework. Kunming was scheduled to host COP 15 for the Convention on Biological Diversity in 2020, but it had to be deferred due to Covid-19. The city eventually handed over the duty of hosting COP 15 in December 2022 to Canada's Montreal. Just slightly over two weeks into my job, I was getting reports on the conclusion of the convention from the civil servants who were sent there. We promptly came up with an updated policy on biodiversity to align it with the global framework.

Wildlife Rangers

The issue of military veterans who leave the service without pension in their early forties is a longstanding one. I am familiar with the issue, as my constituency, Setiawangsa, is home to the Ministry of Defence headquarters with a significant number of military voters, not to mention veterans who continue to stay in the constituency after retirement. Now, a programme suggested by the wildlife department, the community rangers, provides an interesting way to partly address this issue. Through this scheme, rangers are appointed from military veterans and police retirees, as well as indigenous Orang Asli and local communities. Implemented since 2019, this Biodiversity Protection and Patrolling Programme was identified as one of the important measures to increase boots on the ground to protect the Malayan Tigers. Local NGOs have also been brought in.

But a bigger involvement has come from the Orang Asli, who make up more than half of the 1,000 community rangers in Peninsular Malaysia in 2023. Guarding the forest is, after all, a traditional role of the indigenous. This follows the model of the Senoi Praaq, the special Orang Asli unit of the Royal Malaysian Police. It was formed during the Malayan Emergency to utilize their skills in the jungle against the Communists.

During my trip to the Royal Belum, I was introduced to the Menraq—a wildlife patrol protection team consisting of the local Jahai Orang Asli.[100] I asked a few of them about the time they spend in the jungle. Their excursions can last for weeks. Sometimes, they wait the entire day hiding on top of the trees to catch signs of poachers and loggers. To put things into perspective, some of the poachers from Indochina would hide in the jungle for three to four months to place snares. I tried on a backpack for their two-week-long excursions, weighing almost 30 kilogrammes! I was only able to carry it for a few seconds. I could not imagine how they could hike through the challenging trails for hours at a time with that weight on their backs.

Image 21: Meeting the Menraq, the indigenous Orang Asli community-based wildlife protection team, who have received international awards. Through NGO-run initiatives, such as the Menraq, or the government's Community Rangers Programme involving the Orang Asli, military, and police veterans as well as community members, we have started to reclaim more of our forests from poachers, helping the endangered Malayan tiger.

---

[100] 'Menraq' means 'people' in the Jahai language.

The Menraq won the prestigious IUCN WCPA International Ranger Award for the community-led conservation of tigers. It took a while for them to process the significance of the award, but Rimau, the NGO that runs Menraq, are thinking of conducting a training programme by the beach because most of them have never been to one.[101] They are taught to use GPS devices and maps, but their indigenous knowledge of a hunter–gatherer society makes them unique. I have always believed that the role of the Orang Asli in sustainability efforts needs to be recognized. The Orang Asli need to be actively engaged by the government to improve conservation work. In the beginning, the Perak State Park Corporation and Rimau had the challenging task of convincing the *tok batins*[102] to approve the Menraq initiative. Many resisted at first. To win them over, the State Park Corporation and NGO started educational programmes for the villages. Other than stipends for the rangers, there is also a contribution to the community. Additionally, for the first time, this initiative has been expanded to Sabah and Sarawak, where 500 rangers were appointed. From 2019 to the first half of 2023, the value of items seized from poachers has been over RM112 million, and 1,875 snares have been discovered and destroyed.

There are many examples across the world of the significance of indigenous communities in protecting forests. In the Brazilian Amazon, the Baniwa people are renowned for their role in conserving the Alto Rio Negro territory. They believe that even rocks and rivers are living creatures. Thus, human beings have a role to play in protecting the forest. The problem with modern society is that more often than not, everything is reduced to dollars

---

[101] Aqilah, Ili, 'People Who Protect', *The Star*, August 15, 2023, https://www.thestar.com.my/news/nation/2023/08/15/interactive-people-who-protect [Accessed February 20, 2024].

[102] 'Tok batin' means 'village chiefs'.

and cents, but the value of indigenous traditions in protecting wild habitats cannot be quantified.

Green Palm Oil

It is impossible to write a book about the conservation of forests in Malaysia without wading into the palm oil debate. I remember stopping by the Dubai airport duty free shop to buy chocolates for my family and staff after COP 28. I was surprised to see no-palm-oil labels on many products. 38 per cent of all vegetable oils in the world come from palm oil. Originating in Africa, it has been used by humans for over 5,000 years. The use of palm oil grew during the Industrial Revolution. It was first introduced to British Malaya in the nineteenth century but only planted commercially in 1917 in Selangor.

In 1960s, palm oil planting grew rapidly in Malaysia as the government pushed for new sources of income beyond rubber and tin. Malaysia was highly dependent on the two commodities, and any fall in prices majorly impacted the economy. In addition to the major plantation companies, which were mostly owned by the British and some by Chinese Malaysians, the government started FELDA. Abdul Razak Hussein, the first deputy prime minister of independent Malaya and then the second prime minister of Malaysia envisioned a programme to allow for 'land for the landless and jobs for the jobless'. FELDA was a massive rural poverty eradication programme under which the poor were moved to new settlements and started rubber and palm oil smallholdings. Between 1958 to 1990, the last year when new settlers were accepted, more than 112,000 settlers benefited from the programme. The programme received global accolades in eradicating poverty.

Between the late 1970s and early 1980s, three big British plantation companies, Sime Darby, Guthrie, and Golden Hope were acquired by Malaysian interests. This was based on a desire for Malaysia to take back much of the assets that remained in colonial hands after Independence. Malaysia became the world's largest producer of palm oil, remaining an important player even after being overtaken by Indonesia in 2006. Today, Malaysia and Indonesia produce most of the palm oil in the world.

In the 1990s, the use of palm oil grew in food as the public became aware of how unhealthy trans fats are. Palm oil started being used as the base for margarine and as a healthier and cheaper alternative to butter for pastries. It is used in the production of chocolates. But it is also widely used in soaps and shampoos as a foaming agent. At the same time, palm oil is present in lipsticks and lotions along with animal feed. It is estimated that almost half of the packaged items sold in grocery shops contain palm oil.[103] Palm oil is also used in biodiesel, which reduces the use of petrol-based diesel, thus reducing carbon emissions. Its versatility makes it attractive to planters and manufacturers alike.

Much has been said about how palm oil has contributed to deforestation in the tropics, particularly in Malaysia and Indonesia. Yes, undeniably that is a problem. It is important, however, to establish a few facts. Palm oil is the *most efficient* vegetable oil to be planted. Although it supplies almost 40 per cent of vegetable oil globally, it occupies only 5 per cent of the total farmland used for vegetable oils. To switch to alternatives would require up to ten times more land.

That is why, instead of a total unilateral boycott of palm oil (as promoted by the manufacturers of the chocolates I saw in Dubai), many credible international organizations—WWF, Greenpeace, IUCN, and Solidaridad—have all called for sustainable cultivation and production of palm oil. The enemy is not all palm oil but *dirty*

---

[103] Spinks, Rosie, 'Why Does Palm Oil Still Dominate the Supermarket Shelves?', *The Guardian*, December 17, 2014, https://www.theguardian.com/sustainable-business/2014/dec/17/palm-oil-sustainability-developing-countries [Accessed October 1, 2023].

palm oil. A ban will either result in the use of other inefficient vegetable oils or for palm oil producers to look for markets without sustainability standards.

Deforestation from palm oil has also significantly dropped in Malaysia, Indonesia, and Papua New Guinea. WWF, the World Resources Institute, and Chain Reaction Research have all recorded that deforestation linked to palm oil has dropped significantly in 2022.[104]

In Sabah, the Rhino and Forest Fund restores land that has been converted to oil palm plantations back into its original state. It not only functions as a restored forest eventually, but already is a wildlife corridor. Elephants, orangutans, sun bears, and leopards all use the corridors. The plan is to reforest 7,000 hectares of such land to link the Kulamba and Tabin wildlife reserves as well as mangrove forests.

Sustainable Forestry

The Chini lake is the second largest natural freshwater lake in the Peninsula, consisting of twelve smaller lakes connected to one another. The Jakun Orang Asli tribe inhabit the forests. In the middle of the year, the lake transforms as thousands of white and pink lotus flowers blossom. The residents could drink the water straight from the lake in the past.

---

[104] 'The Chain: Deforestation Driven by Oil Palm Falls to a Four-Year Low', Chain Reaction Research, March 7, 2022, https://chainreactionresearch.com/the-chain-deforestation-driven-by-oil-palm-falls-to-a-four-year-low/ [Accessed October 10, 2023]; 'Deforestation Fronts: Drivers and Responses in a Changing World', WWF, 2021, https://wwfint.awsassets.panda.org/downloads/deforestation_fronts___drivers_and_responses_in_a_changing_world___full_report_1.pdf [Accessed October 10, 2023]; Weisse, Mikaela, Liz Goldman, and Sarah Carter, 'Tropical Primary Forest Loss Worsened in 2022, Despite International Commitments to End Deforestation', *Global Forest Watch*, June 27, 2023, https://www.globalforestwatch.org/blog/data-and-research/global-tree-cover-loss-data-2022/ [Accessed July 22, 2023].

However, for many years, the residents have faced numerous problems. Parts of the forests have been cleared. Mines and plantations have also encroached into the area surrounding the lake. Silting as well as leaching of chemicals from mines has polluted the lake, and the beautiful lotus flowers have disappeared. The community living near the lake has been complaining that the water is contaminated.

In a change of fortune, in 2023, they were encouraged to see small clusters of lotus flowers blossoming again in the lake. This was due to the conservation efforts of FRIM and the National University of Malaysia, where lotus seeds were planted on land and then transferred to the lake.

The Matang Mangrove Forest Reserve, Perak, has been systematically managed for over a century, since 1902. Sized slightly over 40,000 hectares, it is internationally considered one of the best examples of a managed mangrove reserve. The trees are managed on a thirty-year rotation. Thinning is done twice in fifteen- and twenty-year blocks. Aside from harvesting mangrove poles, thinning also enables the trees to grow better.[105] Under the Matang management plan for 2010–2019, the reserve has been divided into four zones: 'protective', 'productive', 'restrictive productive', and 'unproductive' forests. There still are virgin forests inside that have virtually been untouched since it was gazetted as a forest reserve. The state forestry department allocates a minimum of 2.2 hectares of productive forests to each of the mangrove pole and charcoal contractors every year.

Other than producing a raw supply for charcoal, the forest is maintained as a nature-based solution or green infrastructure to

---

[105] Goessens, Arnaud, Behara Satyanarayana, Tom Van der Stocken, Melissa Quispe Zuniga, Husain Mohd-Lokman, Ibrahim Sulong, et al., 'Is Matang Mangrove Forest in Malaysia Sustainably Rejuvenating after More than a Century of Conservation and Harvesting Management?', *PLOS One* vol. 9, 8. August 21, 2014, e105069. https://doi.org/10.1371/journal.pone.0105069 [Accessed September 25, 2023].

prevent coastal erosion. The mangrove forest also offers a space for habitat protection, research, and ecotourism. On one school holiday, I decided to take my family to explore the forest, witness the variety of wildlife, and learn about the different trees. If you're lucky, you might even see dolphins. Unfortunately, the day I was there was not one of them. Because Matang lacks the diversity of natural mangrove forests, it raises concerns of lower resilience and inhibits its ecological service. Nevertheless, the example of cooperation between the authorities, entrepreneurs, and local community in sustainably managing the mangrove forest for over a hundred years is one we can not only try to replicate in other parts of Malaysia but even something the world can learn from.

Another well-established success story is the FRIM, the world's largest recreated rainforest. It was established in Kepong, Kuala Lumpur, in 1925, on a barren plot of 1,500 acres. Previously vegetable farming and mining had been done on the land. Various forms of experimentation was done for reforestation, and there were many questions about whether it can be done. However, it was successfully done within a lifetime.

In the 1970s, while hiking in FRIM, people noticed how the canopy of the trees above formed perfect puzzles. The tree canopies covered the forest in its entirety but maintained a sort of 'social distancing' from one another without any overlap or touching whatsoever. If seen from above, the trees look like massive cauliflowers. This phenomenon is called 'crown shyness', as the shoots of trees are sensitive. Another theory is that it prevents leaf-eating insect larvae from infecting other trees. One of the reasons this is so noticeable in FRIM is because this is a reforested area, and the trees were planted around the same time. In 2024, I took the Climate Club for Foreign Ambassadors, a group of diplomats with an interest in climate change, to FRIM to take a look at the forest.

In 2019, the Federal Government launched the EFT worth RM60 million. The money was distributed to state governments to assist them in conservation initiatives. It was increased to RM70 million a year in 2021 and 2022. When the new administration came in, before I could ask for an increase, Prime Minister and Minister of Finance Anwar Ibrahim more than doubled the allocation to RM150 million a year. In 2024, it was increased to RM200 million. The calculations are based on the size of protected areas and their performance—seeing what state governments are doing to protect, rehabilitate, and restore forests. Currently, greater emphasis is given to quantity, i.e. the size of protected areas, but the target is to slowly increase quality, i.e. performance on conservation to be of equal value with the former by 2030.

While many states pushed for even more compensation, we were clear that the EFT was an *incentive*, not a compensation. It helps state governments appreciate that their forest resources have value and must be conserved. It also reduces the Federal–state tensions on forest-related matters. The implementation of this policy in Malaysia is modelled on the EFT implemented in Portugal, France, India, China, and Brazil. For Portugal and France, the policy is coordinated at the national level looking at conservation areas. In India and China, the funds are transferred to specific regions. For Brazil, the states transfer funds to the municipalities considering the conservation area, indigenous land, and other factors. Looking at these examples, the UNDP collaborated with the Malaysian government to come up with the EFT scheme.

Malaysia has also been participating actively in ensuring that forest extraction is sustainable. In Peninsular Malaysia, there is a forest management certification based on the international PEFC. For some years, the government has provided compensation for death and injury due to conflict with wildlife. Realizing that we needed to expand our approach for people who are experiencing increasing human–wildlife conflict due to habitat loss to buy into the programme, we managed to successfully lobby the Ministry of Finance to launch a fund for the indigenous and rural communities for damage to property and crop due to wildlife in the 2024 Budget.

Costa Rica has shown the world how to successfully pursue tropical reforestation. Suffering massive deforestation throughout the twentieth century, by the mid-1980s, only about a fifth of the country's landmass was still forested. Aggressive reforestation policies implemented since then have allowed the forest size to increase three-fold, to 60 per cent. Farmers who practice sustainability benefit from payments for ecological services, largely funded by taxes on fossil fuels. No one can clear forests without approval from the authorities. Landowners can practice selective logging in forests on their plot. This, combined with a thriving ecotourism sector, ensures the local communities truly appreciate the value of forests, not only for the environment but as a source of income.

The Sahara is known today as the world's largest (hot) desert. It was not always this way. The area was lush, wet, and green after the Ice Age. But, more than 5,000 years ago, it began to get dryer. It is speculated that this happened due to changes in our planet's orbit as well as overgrazing by humans. Today, it is observed that the Sahara is growing bigger, taking up 10 per cent more land since records first started being kept over a century ago.

In 2007, the Great Green Wall vision was proposed by the African Union. It is 15 kilometres wide and spans 7,775 kilometres from Djibouti to the east to Senegal to the west. Initially envisioned as a mega tree-planting exercise, today the project does not look at a one-size-fits-all approach. The eleven participating countries are allowed to address the challenge through flexible approaches to consider ways to protect forests and biodiversity and deal with the loss of productive capacity of land. Climate change has disproportionately affected Africa (even though it produces little emissions), and the wall is a plan to mitigate it. The idea is that it will in turn combat poverty, migration, and even terrorism. While progress is still slow, animals are returning to many areas. Some of the trees planted have offered economic opportunities to the local communities. This is crucial because, otherwise, it is likely that the trees will just be cut for firewood.

One of the successful examples of involving local communities is Senegal. The forested land is owned by the communities, instead of the government, giving them a direct stake in the project. Trees are planted in orchards totalling 50,000 acres in size and either produce gum arabic, which has value as a commodity, or fruits directly benefitting local communities.

Imran Khan inaugurated the Plant for Pakistan programme during his tenure as prime minister in 2018. It was modelled on an initiative in Khyber Pakhtunkhwa, when Khan's party was governing it. As part of the provincial programme, the forests across almost 350,000 hectares were successfully restored and a billion trees were planted in the Hindu Kush mountains. This was the first project to meet the target of the Bonn Challenge, an initiative that intends to restore 150 million hectares of degraded forest by 2020. It may be surprising to know that the country is estimated to have the greatest number of glaciers outside the polar areas in the world. But it is now under threat as Pakistan is among the worst countries to feel the brunt of climate change.[106] During the pandemic, 60,000 workers that lost their jobs were hired under the programme.

Before this, we discussed the Water Waqf initiative in Malaysia. In Indonesia, the world's biggest Muslim country, the endowment instrument has been used for forest conservation. Three waqf

---

[106] Hutt, Rosamond, 'Pakistan Has Planted Over a Billion Trees', *World Economic Forum*, July 2, 2018, https://www.weforum.org/agenda/2018/07/pakistan-s-billion-tree-tsunami-is-astonishing/ [Accessed January 2, 2024].

forests have been established in Leuweung Sabilulungan, West Java; Cibunian, Bogor; and Jantho, Aceh. Output is measured in the form of providing environmental services—like water supply, carbon sinks, and energy security—or in the form of education and research for institutions. Forested land that has not been gazetted as a reserve, can be put under waqf to ensure that it is no longer allocated for other purposes. Degraded forests can also be included to be restored.

Challenge of Forest Plantations

The Malaysian government, for its part, introduced forest plantations in permanent forest reserves in an effort to ensure the supply of timber. The idea is to do it on idle land owned by state agencies or on degraded forests and plant fast-growing trees. By doing so, logically, there should be less pressure on natural, pristine forests. The idea was formalized in 2012.

Unfortunately, in practice, it was discovered that over 100,000 hectares of natural forests have been cut down without replanting. This comes as no surprise because replanting is not only expensive but also a long time passes before the investor can harvest their trees. When replanting happens, there is a shift from the diversity of natural forests to a monoculture environment where only one species is grown. Rubber trees are popular, as after a few years, the planters can tap latex for another few years before cutting down the trees. Many plantations do not adhere to the EIA requirements imposed by the authorities.

In 2021, the government decided to impose a fifteen-year moratorium on new forest plantations in permanent forest reserves to evaluate their long-term impact on the environment. In the following year, amendments were made to the National Forestry Act, wherein the maximum fine for occupying or carrying out activities in permanent forest reserves was increased from RM50,000 to RM5 million, while imprisonment was increased from five years to twenty years. Any degazetting of permanent forest reserves must go through a process of public inquiry and must be replaced with at least an equal-sized plot of forest. However, in order for these amendments to take effect, state governments must follow through with their own amendments in state legislation, as forests fall under state jurisdiction.

Protecting our Oceans and Seas

Tioman Island lies off the east coast of Peninsular Malaysia, in the South China Sea. Legend has it that a princess, Seri Gumum, was cursed to be imprisoned in a forbidden garden over a lake. A foreign prince, Seri Kemboja, came and fell in love with the princess, and broke her free from the garden. But the moment they left the garden, both turned into *nagas* or serpents. The forbidden garden flooded and turned into Chini lake mentioned earlier. The two serpent lovers swam onto the river, attracting thunder and lightning as they made their way out to the sea. However, in the sea, Seri Gumum and Seri Kemboja turned into the Tioman and Lingga Islands respectively.

Home to a small population of 3,000 people, much of Tioman Island is covered by forests and is, in fact, a wildlife reserve. The sea is surrounded by lush coral reefs and is designated as a marine park, making the island a popular destination for scuba divers, snorkelers, and surfers. The existing airport had many limitations. The landing strip is short. Airplanes can only arrive and depart in one direction. The size of airplanes that can land is also small. This pushes up the cost of flights.

In 2003, a private corporation announced a plan to build a new airport by reclaiming the sea. This would allow larger planes to land there and increase the volume of passengers while reducing the costs-per-passenger. The reclamation work, dredging, and blasting would have an adverse impact on the marine park. So, the DOE objected. The prime minister at the time, Dr Mahathir Mohamad, claimed that it would not affect the environment.

Five years later, the transport ministry cancelled the project, announcing that environmental concerns as one of the factors contributing to the cancellation.

The proposal was revived in the 2018 Budget. It came to my table in 2023 as the EIA report went on display. 76 per cent of the airport would be on reclaimed land. The new airport would be able to handle Airbus A321s and Boeing 737s and bring flights from as far as eastern India, central China, and western Australia. In its report, the EIA listed physical damage, destruction of habitat for marine life, and disruption of coral photosynthesis due to artificial lighting as demerits of the airport. One of the measures proposed to mitigate the effect was to move the corals—the major attraction to Tioman—to a different location.

I received numerous messages on my WhatsApp, Instagram, and Twitter by Malaysians protesting the project. I brought it up in Cabinet and explained the full ramifications of building a new airport. Finally, the Cabinet decided to not proceed with building the new airport with the impact on the marine park being one of the primary considerations.

Ultimately, climate change is impacting the oceans and seas in a major way. Record-breaking temperatures are actually *cooking* baby turtles. The turtles lay their eggs in the sand. Their eggs dry up as the moisture dissipates, and when the eggs manage to hatch, the baby turtles are not able to get out alive through the burning sand. Hotter temperatures also mean more female babies are produced, affecting the gender mix and reproduction of the turtles. In 2022, Bette Zirkelbach, manager of the Turtle Hospital in Florida, said that because the past four summers were the hottest summers on record (2023 turned out to be even hotter), no male sea turtles had been found.

Oceans are the biggest heat sinks and major carbon sinks on the planet. Ocean temperatures are getting warmer; sea levels are rising while ocean acidification is increasing as it absorbs more carbon dioxide. This means corals and molluscs have less carbonate for their shells and skeletons. Corals look like rocks but are part of the animal kingdom. They play an important role in the marine ecosystem. Along the coast, as elaborated earlier, mangroves have an important and well-understood function in moderating the impact of tides and waves. Healthy coral reefs absorb most of the wave energy and severely reduce coastal erosion.[107]

When you dive near a healthy coral reef, you may hear unique sounds—some describe it as sizzling meat being grilled, others as the snap, crackle, and pop sounds of Rice Krispies in milk, or the crackle of a fire. That is the 'music' of the diverse creatures living in the corals: snapping shrimps producing bubbles, clown fish banging their teeth, or even fish communicating during breeding and fighting. The disappearance of these sounds is a reliable indicator that the corals are dying.

As a result of rising sea temperatures, corals experience shock and turn white. This is because the corals expel the algae that gives them colour. But it is not just the loss of colour that's worrying. These corals become vulnerable to disease and many eventually die. During the El Niño phase from 2014–2017, 70 per cent of corals across the world experienced coral bleaching. This, in turn, leads to a drop in the number of fish and other marine life that depend heavily on corals, affecting livelihoods of communities earning their living from the sea. Since 1950, the world has lost half of its corals.

However, due to the enormous amount of carbon already absorbed by the oceans, their function as a carbon sink is reduced, meaning more carbon is left behind in the atmosphere.

The reef ball artificial coral technology was invented by an American, Todd Barber. It can be floated and transported by boats

---

[107] Ferrario, Filippo, Michael W. Beck, Curt D. Storlazzi, Fiorenza Micheli, Christine C. Shepard, and Laura Airoldi, 'The Effectiveness of Coral Reefs for Coastal Hazard Risk Reduction and Adaptation', *Nature Communications*, vol. 5, 3794. May 13, 2014, https://doi.org/10.1038/ncomms4794 [Accessed September 13, 2023].

to sea. The environmentally friendly reef balls attract algae, coral, and sponges. The holes that line it allow fish to shelter inside. They also have sharp surfaces that tear trawler nets. Sarawak became the first place in Asia where reef balls were introduced at the end of the 1990s. Now, nearly 17,000 reef balls have been installed there, stretching across 720 kilometres of the state's coastline along the South China Sea. The artificial reefs have positively affected the local fishing community.

On Hatamin Island, Indonesia, the Coral Guardian organisation has initiated participatory marine conservation. Some of the sites were ravaged by dynamite fishing. They interviewed fishermen devastated by the disruption to their livelihoods. Coral Guardian replants pieces of corals that have been fragmented but are still alive. This can be done on a solid structure or directly onto the seabed, depending on the local condition.

Forests and gardens epitomize paradise in various faiths and religions. In Islam, *Jannah*, or paradise, is described as gardens filled with lush trees, beautiful flowers, dotted with springs and rivers flowing underneath. In fact, the word 'paradise' in English comes from the Old Persian *pairidaeza*, which described walled gardens. The highest level of paradise in Islam, *firdaus* (which came from the same Old Persian word), is interpreted as being on a mountain at the centre of Jannah. In the Abrahamic religions of Judaism, Christianity, and Islam, Adam and Eve originally were situated in the Garden of Eden, surrounding the Tree of Life.

For billions of years after the earth was formed, there was hardly any life on it. The atmosphere was what we fear our current atmosphere will be if we do not address climate change: it was hot with volcanoes spewing carbon dioxide into the air (even so, the carbon dioxide being emitted by human activities today is many times that of what is being emitted by volcanoes!). The ozone layer was too thin to filter UV rays from the sun. Between 400 to 500 million years ago, plants from the sea began to colonize dry and barren land. They adapted to living in drier environments.

Plants then began to expand aggressively, reducing the amount of carbon dioxide while producing oxygen.

The planet needs our forests and oceans to function. It needs wildlife living within the forests and oceans to sustain our existence. We have spoken about the energy transition and sustainable use of water above, but without preserving our massive carbon sinks, we will not be able to achieve our goal of leaving behind a liveable planet for our children. Malaysia has a unique role to play in this fight due to our inherent advantages.

In 2021, the UK government published a 600-page study that looked at the failure of economics in internalizing the cost of loss of biodiversity, titled *The Dasgupta Review*. The economy, essentially, is embedded in, not external to, nature. Nature is an asset, yet the value it provides goes beyond economic value alone. The world has reached this stage not only because of failures of the market but also of institutions. Biodiversity makes nature resilient to disruptions. Much of the planet is either at the tipping point or has passed it altogether. This makes taking action all the more urgent.[108]

Mark Carney, who was governor of the Bank of Canada and then governor of the Bank of England, pointed out this irony in his seminal book, *Value(s)*:

> Why is Amazon rated as the world's most valuable companies by financial markets, but the value of the vast geographical region of the Amazon rainforest appears on no ledger until it is stripped of its foliage and converted into farmland?[109]

---

[108] Dasgupta, Partha, 'The Economics of Biodiversity: The Dasgupta Review', *HM Treasury*, 2021, https://assets.publishing.service.gov.uk/media/602e92b2e90e07660f807b47/The_Economics_of_Biodiversity_The_Dasgupta_Review_Full_Report.pdf [Accessed February 20, 2024].

[109] Carney, Mark, *Value(s): Climate, Credit, Covid and How We Focus on What Matters*. Revised and updated edition. London: William Collins, 2021, p. 15.

Climate Justice for All

> 'History belongs in the past; but understanding
> it is the duty of the present.'
>
> —Shashi Tharoor, Indian politician and writer

Climate change, environmental issues, and forest conservation—all these may seem like abstract concepts, compared to the bread-and-butter issues that often dominate the minds of the public (and thus, politicians). Indeed, as a backbencher and opposition MP, as well as a senior party leader prior to becoming minister, I normally zoomed in on wages, jobs, and cost of living, as these issues grabbed the attention of my constituents as well as voters in general. In addition—not just in Malaysia but in many parts of the developed world as well—we have identity politics and culture wars that are waged by demagogues to energize their base and divide society. In Malaysia, this takes the shape of racial and religious politics. Most dangerously, such politics often seek to exploit the frustrations of those being left behind: the poor and the marginalized.

Globally, as the cost of living escalates, many politicians are tempted to give environmental and energy transition policies mere lip service, just as many corporates tend to greenwash. Worse, some portray climate action as expensive distractions promoted by an out-of-touch elite that contributes to the cost-of-living crisis. The far-right in Europe and Donald Trump in the US have gained support by declaring that climate change is fake news or, at the very least, is being exaggerated by the establishment and the mainstream media. I receive criticisms on my social media that my pursuit of addressing the planetary crises means that I am pandering to Western liberals or that it is a luxury Malaysia cannot afford to undertake. On the contrary, protecting the planet is about our survival. It is a necessity.

Diplomacy, Protest, and Natural Resources

The second half of the twentieth century saw oil taking centre stage in shaping international geopolitics. Its influence arguably extended to the sustainability (or lack of) of democratic institutions in some countries, the willingness of superpowers to go to war, and how economic policies were designed by petrol politics.[110] In this century, water will shape international relations, where failures in diplomacy can cause conflict. As mentioned earlier, the shadow of the tensions with regards to Johor supplying water to Singapore has loomed large even in the discussion on Malaysia exporting renewable energy to our southern neighbour.

In fact, tensions frequently escalate between different states and provinces on the use of water *within* national borders. In Malaysia, two states to the north—Perlis and Penang—depend on water from the Muda river basin in Kedah. Penang is in discussions with Perak about diversifying their source of water because the state's demand, as an industrial and commercial one, is high while they do not have a sufficient catchment area. Malaysia's richest state, Selangor, provides water not only to its own residents but also to Kuala Lumpur, the capital and commercial centre of the country, as well as Putrajaya, the administrative centre. However, Selangor does not have sufficient water and supplements their water from Pahang.

---

[110] Mitchell, Timothy, *Carbon Democracy in the Age Oil: Political Power in the Age of Oil*. London: Verso, 2023.

While most people know about the Indian freedom fighter and icon of non-violence Mahatma Gandhi, Abdul Ghaffar Khan (also known as Bacha Khan) is less well-known. His other moniker is Sarhadi Gandhi or 'the Frontier Gandhi'. He started a non-violent movement against the British rulers of India using principles drawn from Islam. Abdul Ghaffar was against the partition of India, but after the formation of Pakistan, pledged his allegiance to the country. However, conflicts with the Pakistani government meant that just as he had done during the British era, he continued to spend many years in prison. In the early 1960s, he was named Amnesty International's Prisoner of the Year and around two decades later, he was nominated for the Nobel Peace Prize. But it is his last major political intervention prior to his death that is significant—he stopped the construction of the Kalabagh hydroelectric dam. Tensions between how different provinces would benefit from the construction of the dam at the environmental expense of other provinces were at the heart of the dispute.

In Los Palacios y Villafranca, Spain, farmers switched from their traditional party, the Socialist Workers Party, to the far-right Vox in 2022. After a very dry and warm season, Vox attacked the EU's rules banning irrigation from the Doñana wetlands, a UNESCO World Heritage Site. It struck a chord with the impoverished farmers. Water theft has become a serious issue in Spain due to depleting water supplies. Prior to that, in 2018, Priscilla Ludosky, a cosmetic entrepreneur in France, started a campaign on Change.org to push for lower petrol prices. The number of supporters started to grow, and Eric Drouet, a lorry driver, got involved. The Yellow Vests Protests, or *Mouvement des Gilets Jaunes*, erupted in France due to a planned increase in fuel taxes for diesel prescribed by neoliberal economists. The protesters marched wearing high visibility jackets that were synonymous with lorry drivers.

Five years later, the German coalition government led by Olaf Scholz attempted to pioneer an ambitious law on heating for buildings. Scholz comes from the left-of-centre Social Democrats but includes the environmentalist Alliance 90/the Greens as well as the pro-business Free Democratic Party. The legislation will basically ban all oil and gas heating systems and replace them with heat pumps that are more efficient. It concentrates surrounding heat from outside the property to warm the insides. It also runs on electricity, and, on the back of more renewable energy, will be crucial in the journey towards net-zero. But each unit can cost up to EUR20,000 more than the traditional gas boiler. At this point in time, it is bigger than gas boilers, must have extra insulation, and needs outdoor space. Even with the authorities promising to step in to partially fund the cost for poorer households, strong opposition remains.

Some countries are advocating to adopt net-zero goals that may be too ambitious to be achieved by 2025 or 2030. British environmental economist Dieter Helm has warned such countries that this 'medicine' may kill the patient. Costs of living are likely to increase dramatically as the price of pollution gets internalized, industries may collapse, and people may lose their jobs. He reminded us of the doctrinaire, out-of-touch economists who prescribed shock-therapy in former Communist countries in Eastern Europe and the former Soviet Union. Prices of staple food that were heavily subsidized during the Communist era rocketed up, making the ordinary people demand the return of the old authoritarian policies.[111]

In developed economies across the world, inflation has been increasing in the post-pandemic era. There have been arguments that as the world embarks on energy transition and internalizes the price of carbon, costs for businesses will go up. Carbon taxes are an important tool to achieve net-zero, but they add to the costs of fossil fuels. Another reason behind the rising prices of

---

[111] Helm, Dieter, *Net-zero: How We Stop Causing Climate Change*. London: William Collins, 2021, p. 93.

fuels is that investments in fossil-fuel production are drying up in many countries. Traditionally, an increase in price gets companies to invest more, which increases supply and brings down prices. Today, however, that action is shunned by investment funds, activists, and even certain governments.

Historically, the cost of producing fossil fuels tends to go up, whereas cost of producing renewable energy goes down.[112] The fuel for the latter, by definition, is constantly replenished. Fossil fuels, too, are largely stored solar energy, but they take millions of years to form. Presently, our consumption of fossil fuel far exceeds the rate of its formation, rendering the resource difficult to be replenish and environmentally unsustainable. On the other hand, the cost of producing electricity from large-scale solar farms has decreased by 88 per cent in just eleven years from 2010 to 2021. During the same period, the cost of producing electricity by harnessing the wind has been reduced onshore by 68 per cent and offshore by 60 per cent. The pattern is similar to what we see in electronic and digital goods, where prices go down for products that are even more powerful than before.[113] Prices of battery packs for EVs dropped 89 per cent between 2008 to 2022.[114] But here, assistance and even reparation from developed

---

[112] 'Is a Sustainable World an Inflationary World? Part 1 of 3: The Net-zero Question', Generation Investment Management LLP, 2022, https://www.generationim.com/our-thinking/insights/is-a-sustainable-world-an-inflationary-world-part-1-of-3-the-net-zero-question/ [Accessed August 18, 2023].

[113] 'Renewable Power Generation Costs 2021', IRENA, 2022, https://www.irena.org/publications/2022/Jul/Renewable-Power-Generation-Costs-in-2021 [Accessed August 18, 2023].

[114] 'FOTW #1272, January 9, 2023: Electric Vehicle Battery Pack Costs in 2022 Are Nearly 90% Lower than in 2008, according to DOE Estimates', Office of Energy Efficiency and Renewable Energy, January 9, 2023, https://www.energy.gov/eere/vehicles/articles/fotw-1272-january-9-2023-electric-vehicle-battery-pack-costs-2022-are-nearly [Accessed August 18, 2023].

countries become crucial, as it is the developing world that feels the price crunch when fossil fuels become more expensive.

We also need to recognize that one of the causes of our climate and environmental crisis today is our unsustainable consumption. Hence, we need to change our lifestyle if we are to achieve our net-zero goals. As much as politicians like me would prefer it, there is no such thing as a free lunch. The transition will have some costs, as goods that have negative externalities to the environment are priced higher.

Young philosopher William MacAskill argues for long-termism: our moral priority should be about positively influencing the long-term future. As mentioned in the beginning of the book, the problem is that the media tends to operate on a 24-hour news cycle while elected officials work in five-year election cycles. In fact, our choices should not be merely limited to our generation or our children or their children but even beyond. Theoretically, we should be planning and working for the rest of mankind that will come after us. Humanity, MacAskill argues, is like a teenager. Our species has a long future ahead of us. Just like a teenager who has the choice to focus on school or on playing truant and breaking the law, we must decide between instant or delayed gratification that will have far-reaching consequences for our future as a species.[115] As all of us who have been teenagers know, instant gratification is very seductive and easy.

In fact, it was only in the 1960s that oil and gas began to be utilized in such an unsustainable manner. It was beyond what was necessary for industrial sectors in the West. More efficient refining processes and the expansion of infrastructure on the back of massive public subsidies during the Second World War propelled the industry's growth. The highway system was growing in America along with the ideal of having a suburban, detached

---

[115] MacAskill, William, *What We Owe the Future*. New York: Basic Books, 2022, p. 6.

house. Cars became bigger and heavier. Urea, a product of natural gas and petroleum, began to be used widely in agriculture. The use of single use plastics expanded exponentially. The industry, however, promoted this as something that happened due to their superior management and innovation. The low price of oil spurred the industry to promote more use of the commodity.[116]

In contrast, the idea of 'Doughnut Economics' began to gain ground to describe an approach to economics that is relevant for the twenty-first century. Developed by economist Kate Raworth, it captures the idea of sustainable development and climate justice aptly using a ring doughnut-shaped diagram. Come to think of it, it should be named *keria* economics, as not all doughnuts are ring-shaped, but all keria (a sweet potato-based local dessert, covered in sugar or syrup) are ring-shaped!

Anyway, the outer ring of the doughnut is the ecological ceiling, while the inner ring is social foundation. The sweet spot is what Raworth calls 'the safe and just space for humanity' within the doughnut, in between the ecological ceiling and social foundation.

The social foundations comprises energy, water, food, income and work, housing, gender equality, and social equity. If this all sounds familiar, well, it's because these ideas overlap in many ways with the seventeen SDGs that were agreed by UNGA in 2015. The economy must be able to deliver these needs. We must acknowledge that a lot of the economic growth in the twenty-first century has contributed to millions and millions of people not falling short of the social foundation. Stable electricity and water supplies enable people to lead productive and healthy lives. Most women, who make up half of society, were not able to get the same education and work opportunities as men previously. While many hurdles remain, women's entrance in schools, universities,

---

[116] Mitchell, Timothy, *Carbon Democracy in the Age Oil: Political Power in the Age of Oil*. London: Verso, 2023, p. vii.

offices, and entrepreneurship in most countries provided another engine for economic growth. The Green Revolution in agriculture fed millions of people, improved incomes for farmers throughout the world, and prevented more land from being used for food production.

The flipside, however, is that more often than not, economic growth exceeds the ecological ceiling when growth is no longer sustainable. Breaching the ecological ceiling results in air pollution, ocean acidification, ozone layer depletion, biodiversity loss, and climate change, among others. If we merely focus on producing more electricity and water, even with more sustainable methods, without accepting that there must be limits to consumption, we will not turn the tide on our planetary crises. Many countries in the developed world have breached the ecological ceiling, while most of the developing world is still struggling to fulfil the basic requirements to meet the social foundations. In fact, as we know, the poor are most susceptible to the impact of climate change, environmental pollution, and biodiversity destruction. As a planet, we need to operate within the doughnut to leave behind a sustainable world for future generations.[117] This is where capitalism falls short. As Mahatma Gandhi famously said, 'The world has enough for everyone's needs, but not everyone's greed.'

People tend to forget that while much focus is given to Adam Smith's *Wealth of Nations* as the foundation of modern capitalism, he actually wrote *The Theory of Moral Sentiments* first, in which he reflected:

> And hence it is, that to feel much for others and little for ourselves, that to restrain our selfish, and to indulge our benevolent affections, constitutes the perfection of human nature; and can alone produce among mankind the harmony

---

[117] Raworth, Kate, *Doughnut Economics: Seven Ways to Think Like a 21st Century Economist*. London: Penguin, 2017, pp. 44–46.

sentiments and passions in which consists their whole grace and propriety.[118]

We do not want the developed world to, in the words of South Korean economist Ha-Joon Chang, 'kick away the ladder'. Chang is referring to how rich countries would tell poorer countries to pursue free market policies while the rich would pursue protectionist and interventionist policies earlier.[119] Similarly, while the rest of the world is willing to pursue more environmentally and climate-friendly policies, advanced economies must assist and facilitate this. We cannot enact climate policies at the expense of the poor. This is the basis of the 'common but differentiated responsibilities' principle that was established in the 1992 Earth Summit.

In 1982, the year I was born, Dr Mahathir Mohamad was just over a year into his premiership (he went on to serve for twenty-two years as prime minister in his first stint, and then I served as a backbencher to Mahathir's twenty-two-months long PH government). He spoke at UNCLOS and Antarctica and warned against countries claiming the continent as their territory. There is no indigenous human population there. Today, around 1,000 (during winter) to 5,000 people (during summer) occupy various research facilities there. Mahathir said:

> A number of countries have in the past sent expeditions [. . . and] have gone on to claim huge wedges of Antarctica for their countries. These countries are not depriving any natives of

---

[118] Smith, Adam, *The Theory of Moral Sentiments*. Stewart edition. London: Henry G. Bohn, 1853, p. 27. https://oll.libertyfund.org/titles/smith-the-theory-of-moral-sentiments-and-on-the-origins-of-languages-stewart-ed [Accessed February 14, 2024].

[119] Ha-Joon, Chang, *Kicking Away the Ladder: Development Strategy in Historical Perspective*. London: Anthem Press, 2002.

their lands. They are therefore not required to decolonise. But the fact still remains that these uninhabited lands do not legally belong to the discoverers as much as the colonial territories do not belong to the colonial powers.

Like the seas and the sea-beds these uninhabited lands belong to the international community. The countries presently claiming them must give them up so that either the United Nations administer these lands or present occupants act as trustees for the nations of the world.[120]

This started a period where Malaysia pursued a diplomatic challenge to the Antarctic Treaty. The agreement, which came into force in 1961, was the first arms control treaty ratified during the Cold War between the US and the Soviet Union, declaring Antarctica as a scientific preserve. Initially, the US wanted to allow atomic testing on the continent, with prior notice to other countries. Argentina, with the support of the Soviets and Chile, managed to push for a total ban on all atomic testing. When there were talks of exploiting minerals there—including by the UK and New Zealand—Mahathir wanted the countries involved to surrender their territorial claims to the UN, which would act as a trustee for all the member states. France and Australia put a stop to the idea of any mining aside from research in Antarctica until 2048, when the treaty will come up for renewal.

Today, the approach has shifted. Malaysia ratified the Antarctic Treaty in 2011, focusing on scientific cooperation instead. The Sultan Mizan Antarctic Research Foundation, an agency under NRECC and later NRES, has been conducting research on polar areas and climate change there since 2012.

---

[120] Mahathir, Mohamad, 'Malaysia National Statement to the 37th UN General Assembly', September 29, 1982.

Subsidies for the Poor

The concept of climate justice underlines that in order to succeed, the problems of the economy and climate must be dealt with together. The basic bread and butter issues are closely related to energy, water, conservation, environment, and climate. Blanket low energy and water tariffs are relevant to the old model of the low-cost and low-value economy that many developing economies started with. Yet, it is like an addictive drug that is hard to overcome, leading to unsustainable use of finite resources. It also leads to the suppression of wages, which is a pivotal part of the vicious cycle of a low-value economy. In fact, perversely, the rich benefit more from blanket subsidies than the poor.

In October 2023, Prime Minister and Finance Minister Anwar Ibrahim estimated that subsidies for fuel, electricity, and food, among others, are expected to reach a total of over RM80 billion for that year. This exceeded the forecasted amount in the budget by a distance, due to rising fuel prices. It is easy to blame the consumers for being wasteful when the economics is not made to work. The 2023 total budget was RM388 billion, which means subsidies stood at over a fifth of government expenditure.

Let us ask ourselves: at the end of the day, why should the rich get subsidized petrol, electricity, and water? The cheap prices for these finite resources leads to wasteful consumption.

Yes, there will be an increase in the cost of living, but we can and must continue to only compensate or subsidize the most deserving. Progressives must not pursue extreme austerity policies by cutting all subsidies and social spending entirely; instead, we should promote *meaningful* subsidies and roadmaps for transition for the urban poor and rural communities. Remember that, in the past, the low-cost economy worked both ways: the blanket subsidies that kept cost of living low justified low wages that were central to the model. As Malaysia's Unity Government moves towards a more value-added economy and a progressive wage model for the public, we must transition from blanket to targeted subsidies as well. We must build a new economic model.

Greener alternatives—such as public transport and electric vehicles, installation of solar photovoltaic panels and greater energy efficiency as well as rainwater harvesting and water efficient products—will only truly take off when the public is not receiving blanket subsidies for petrol, electricity, and water and the right price signal is given to the market.

On the other hand, climate change leads to changing and more intense weather patterns. We suffer heavier or less rainfall than average. We experience floods, rising sea levels, erosion, and more extreme temperatures that all impact the environment and agriculture. Decreasing yields due to the climate crisis lead to higher cost of living. The importance of food security was highlighted during the Covid-19 pandemic. This will be further undermined and, as climate change has a global impact, might lead to reduced supply globally anyway.

I repeat—both within countries or globally, it is the poorest segments of society and the developing economies that bear the brunt of climate change. We are also seeing a growing middle-class emerging throughout developing economies, including Malaysia. They expect to have the same standard of living as their counterparts in the richest countries—and rightly so. The challenge is to see this continue while mitigating its impact on the climate and environment. Abhijit V. Banerjee and Esther Duflo wrote:

> Climate change is massively inequitable. The lion's share of carbon dioxide-equivalent emissions are being generated either in rich countries or to produce what people consume in rich countries. But the greatest share of the cost is, and will be, experienced in poor countries.[121]

---

[121] Banerjee, Abhijit V. and Esther Duflo, *Good Economics for Hard Times*. London: Penguin, 2020, p. 208.

While developing countries are told to cut subsidies to balance budgets and fight climate change, members of the G20—a group of the world's twenty biggest economies—continued subsidizing coal, oil, and gas to the tune of US $1.4 trillion in 2022! This included direct subsidies, investments by state-owned enterprises, and loans from public finance institutions. G20 leaders had pledged to phase out fossil fuel subsidies in 2009. During COP 26 in 2021, global leaders had declared such subsidies 'inefficient' and demanded to ramp up moves to stop them. During COP 28, the world pledged to transition from fossil fuels. At the same time, the energy crisis following the opening of economies after the Covid-19 pandemic and the Russia–Ukraine war pushed up prices, allowing energy companies to earn bumper profits.

The developing world is not asking for handouts, merely fairness. Without a sense of fair play being a central tenet of our policies to deal with the triple planetary crisis of environmental degradation, biodiversity loss, and climate change, we will not be able to come together to collectively find a solution to it.

Fighting Ecofascism

The far-right is often associated with climate denialism. But there is a new strand of thinking that is emerging among them—ecofascism.

In the past, the arguments against immigration were made based on how it allegedly undermines national identities or brings in dangerous ideologies into a country's borders. Today, some argue against immigration on the basis that the influx of foreigners push the environmental limits of a country. In 2021, Arizona's Republican Attorney General filed a lawsuit to reinstate Donald Trump's restrictive immigration policies to overcome 'pollution and stress to natural resources' caused by migrants. Trump, of course, was never a big fan of policies to fight climate change. One American group, Negative Population Growth, calls for the elimination of illegal immigration, quasi-legal immigration (read: refugees), and an 80 per cent reduction of legal immigration. They argue if historically the US was known to be generous as a nation of migrants, today, they should generously reduce its massive carbon footprint.

In 2019, the world was shocked when a twenty-eight-year-old Australian, Brenton Harrison Tarrant, committed two mass-shootings in Christchurch, New Zealand. Taking place in two mosques during Friday prayers, Tarrant livestreamed the attack on Facebook in the style of a first-person shooter video game. Here, however, fifty-one real people died and forty were injured. Tarrant posted a seventy-four-page manifesto voicing his support for the Great Replacement theory, which claims that the global elite is responsible for white people being systematically replaced by non-white people in Europe. He also claimed to be an ecofascist, complaining about the high birth rate of immigrants along with the extreme consumerism of the American capitalist economy.

Although officially targeting overpopulation in general as a stress to the environment and climate, many of the arguments ecofascists make have racist undertones. Then, there is Islam—the fastest-growing religion in the world that has been turned into a bogeyman in the West. After all, when one compares emission per capita, it tends to be the developed and rich nations that top the charts, not developing countries.

Despite all this, we must, at all costs, avoid having a 'we know best' attitude. Many ordinary people want to listen to reasoned arguments. It is difficult to make the case to protect the environment and fight climate change if we keep coming to them from a position of self-righteousness. People want to be *talked to*, not *talked down to*. Otherwise, green politics will always be boxed as an elite, urban movement, and not something that is relevant for everyone in society. Robert Habeck, Germany's minister for economic affairs and climate action as well as former co-leader of Alliance 90/the Greens, lamented:

That's something that dates to the green movement's origins. To survive as a grassroots movement you have to claim to have access to some higher form of truth that others don't. But as we Greens are transitioning to something with a broader political appeal, we are working to reduce that claim to truth and have the better arguments instead.[122]

Most crucially, climate change and environmental disasters in and of themselves are now a major cause of migration—what is called 'climate migration'. As floods wreak havoc on towns and villages, droughts dry up farmland, sea levels rise, and water resources get stressed, millions have to leave their homes for the sake of making a living, but most importantly, simply to *live*. The World Bank estimates that by 2050, there will be more than 140 million internal climate migrants. While, for now, a lot of climate migration happens internally, it is also going across borders.

In May 2023, Cyclone Mocha hit Myanmar and Bangladesh. A million Rohingyas, who suffer from persecution in their homeland in Myanmar, have made the Cox's Bazar refugee camp their home. 450,000 Rohingyas in Cox's Bazar suffered from the cyclone. Homes were destroyed and the refugees were soaked in rain. Their homes are made from extremely flammable materials and, during the scorching afternoons, all it takes is a spark to burn them. At the same time, the growing refugee camp has led to the disappearance of 2,000 hectares of forest. In Mongolia, the excessive number of goats leads to overgrazing, which, in turn, leads to expanding deserts. This brings about what the Central Asians call *dzud* or extreme winter conditions that lead to the

---

[122] Oltermann, Philip, 'Greens Must Shed "Moral Superiority" Image, Says German Vice-Chancellor', *The Guardian*, September 26, 2023, https://www.theguardian.com/environment/2023/sep/26/greens-moral-superiority-german-vice-chancellor-robert-habeck-climate [Accessed October 23, 2023].

death of livestock. In Mexico, droughts lead to villagers leaving rural areas and migrating to the US.[123]

Poor countries, such as Bangladesh and Ethiopia, suffer heavily from domestic climate migration and having to host massive migration from neighbouring countries adds to the environmental stress of the host country.

The Rich Must Be Held Responsible

Malaysia contributes less than 1 per cent to the global greenhouse gas emissions. In contrast, the largest and most developed economies of the world that have benefitted from the Industrial Revolution over the past two centuries—the US, Europe, Japan, and China—are the top emitters. The US is responsible for a quarter of emissions since 1751, the EU 22 per cent, China 12.7 per cent, Russia 6 per cent and Japan 4 per cent.[124] Looking at this from another angle, however, four fifths of the fossil fuels consumed since the dawn of industrialization came about since the 1960s. The US and Canada produce 20 tonnes of greenhouse gasses per capita annually. Other industrialized countries in Europe, East Asia, and Russia (as well as Malaysia) produce, at most, half of that. But even in North America and Europe, the poorest half of the population produces one-sixth of the greenhouse gasses of the wealthiest 10 per cent. Less developed

---

[123] Cantor, David James, 'Cross-Border Displacement, Climate Change and Disasters: Latin America and the Caribbean: Study Prepared for UNHCR and PDD at Request of Governments Participating in the 2014 Brazil Declaration and Plan of Action', *UNHCR*, July 2018, https://www.unhcr.org/media/cross-border-displacement-climate-change-and-disasters-latin-america-and-caribbean [Accessed November 11, 2023].

[124] Ritchie, Hannah, 'Who Has Contributed the Most to Global CO2 Emissions?', *Our World in Data*, October 1, 2019, https://ourworldindata.org/contributed-most-global-co2 [Accessed July 8, 2023].

nations and poorer segments of society are on the course to meet the 2030 targets in line with 1.5 degrees Celsius of warming.[125]

It is ironic, then, when the West tries to play the sheriff by policing the developing world on climate matters instead of constructively supporting the Global South make the necessary transitions. The US was the only signatory of the 1997 Kyoto Protocol that did not ratify it (based on the complaint that there was no commitment from the developing countries). Barack Obama's administration signed the US up to the subsequent 2015 Paris Agreement, which includes both the developed and developing world. However, when Donald Trump took over, the US withdrew, only to re-enter again when Joe Biden became president.

In Malaysia, the big corporations and the richest segments of society have benefited from blanket subsidies. While they lobby the government for more subsidies for renewable energy adoption, cheaper green tariffs and even more incentives for electric vehicles, these will only be achieved at the expense to the majority Malaysians. When California provided massive solar incentives to households, it faced backlash because it meant ordinary Californians were subsidizing the rich to install solar panels.

In fact, there is a small but powerful overlapping class in society: the 'polluter elite'. Oxfam reported that, in 2019, the wealthiest 1 per cent of the world's population produced carbon emissions equivalent to the poorest two thirds of the population—consisting of five billion people in total.[126] In a case

---

[125] Mitchell, Timothy, *Carbon Democracy in the Age Oil: Political Power in the Age of Oil*. London: Verso, 2023, p. vi.

[126] Khalfan, Ashfaq, et al., 'Climate Equality: A Planet for the 99%', Oxford: Oxfam International, 2023, https://policy-practice.oxfam.org/resources/climate-equality-a-planet-for-the-99-621551/ [Accessed November 21, 2023].

of morbid irony, the carbon emissions of the top 1 per cent—who live comfortable, air-conditioned lives—are enough to cause 1 million excess deaths due to heat. And surprise, surprise, those most susceptible to heat-related deaths are the poorest and most marginalized, including migrants, women, and children.

Thus, industrialized nations and the richest segments of society must not just tell the rest what to do but bear their share of responsibility as well to pay for the ecological debt they owe. Industrialized nations should pay a form of climate reparations as financial assistance to the rest of the world. For the wealthiest corporations and richest households in Malaysia, it means paying market rates for utilities and fuel to allow the country's limited resources to be focused on assisting our middle-class and poorest segments of society.

You can't be driving a Porsche or BMW and get to use subsidized fuel simply because you are resorting to RON 95! I must make a confession here. I tried to keep using RON 97 for my car (I did not drive a Porsche, but a Volkswagen Passat), since I first bought it to enjoy its turbocharged engine. The RON 97 price was about RM2.70 per litre when I bought my car. It went down to RM1.60 during Covid-19 and then crossed RM4.00 and nearly hit RM5.00. By the time it crossed RM3.00, I switched to RON 95 because, truthfully, I would rather save money than enjoy a barely noticeable enhanced performance for my car.

Then, as mentioned earlier, in France, Gilets Jaunes protests erupted due to a proposed increase in fuel taxes. Some portrayed the movement as a backlash against climate policies, but the protesters argued that ordinary French households had to bear the bulk of the taxes while carbon-intensive companies were granted many exemptions. The French government had also recently reduced taxes for the wealthiest members of the society. Ordinary French residents had been suffering from a falling disposable income for some time. In short, the protests were not anti-climate action but pro-social justice.[127]

This can only be addressed by international cooperation in promoting climate justice while governments make systemic changes domestically. Without bold and substantive reforms by states and corporations, we will not see the necessary changes to avert our climate disaster.

There have been precedents for what is being advocated here. A major challenge for the world following the Second World War was to rebuild a devastated global economy. The US implemented the ambitious Marshall Plan in Europe and similar policies in Asia as part of post-war reconstruction. Relying on Keynesian economics, they spent US$13.3 billion (equivalent to US$173 billion today) in Europe alone. Rather than focusing on wartime destruction, it looked at the necessary structural reforms to modernize the European economy. Not only did the plan overcome the poverty caused by the war but it also provided the engine that drove the growth of the European economy for over two decades. That era has been dubbed the Golden Age of Capitalism. Debts carried forward by the German government in the 1920s were substantially discounted and stretched over a longer period.

The US had learned from the mistakes that led to the rise of Hitler in Germany. The punitive demands by the victorious Allies for reparations from Germany in the Treaty of Versailles following the First World War led to the economic and social dislocations that made the rise of Nazism possible. Post the Second World War, much of the money transferred through the Marshall Plan allowed Europe countries to rebuild their economies. It was not

---

[127] Bouyé, Mathilde and Yamide Dagnet, 'The Yellow Vests Movement Isn't Anti-Climate Action; It's Pro-Social Justice', *World Resources Institute*, December 7, 2018, https://www.wri.org/insights/yellow-vests-movement-isnt-anti-climate-action-its-pro-social-justice [Accessed August 18, 2023].

purely a selfless decision. Europe, having rebuilt its economy, could use the money to buy American products that benefited the American economy in turn. A similar approach was introduced in Asia to assist Allied adversaries—the Japanese and subsequently the Koreans and Taiwanese—rebuild their economies.

Why shouldn't we have a Green Marshall Plan for our era? After all, many developing countries are also providing an important ecological service to the planet and not just within their borders. This includes mangroves, tropical forests, and the oceans in general. This role, and the crucial need for developing countries to keep these resources intact for the planet, should be compensated. Not only have developed countries been largely responsible for the massive deforestation since the start of the Industrial Revolution within their own boundaries but they have also been largely responsible for the environmental damage and extractive activities that took place in the developing countries when the latter were colonized by the former.

Nowhere is this more urgent than in the small island nations that are feeling the impact of sea-level rise. In 2021, as the Covid-19 pandemic was still disrupting many international events, Tuvalu Foreign Minister Simon Kofe sought to highlight his country's woes by recording his statement to COP 26, speaking in a suit and tie while standing in the sea with his trousers rolled up.

Thus, the discussion for climate reparations—otherwise known as loss and damage funds—should be guided by a sincere attempt to assist developing countries most affected by climate change. During COP 27, the developing world scored a victory when, after a three-decade battle, a loss and damage fund was established. The follow-up discussions, however, did not reach a consensus. The EU attempted to limit the fund mostly to the least-developed countries and small island states. But this would exclude middle-income countries that had suffered major floods such as Pakistan and Libya (and not forgetting, Malaysia).

In November 2023, less than a month before COP 28, where the funding arrangements were supposed to be finalized, the US made a last-minute objection against a word in the text 'urging' developed countries to support the fund.

The US and EU are also pushing for the fund to be controlled by the US -controlled World Bank. This has been strongly protested by the developing world and China. After all, the World Bank is renowned for being sluggish in responding to climate-related crises and natural disasters. But there's also an undeniably political dimension lurking beneath this proposal. A lot of the money provided is in the form of loans, which historically have unnecessarily burdened developing countries. It also enables Western nations to shirk their responsibility towards meaningful climate financing, allowing them to conveniently pass the buck to an institution with a track record of delays and inadequacies. This is why, at COP 28, I argued that the fund's management must involve other international financial institutions—such as the European Investment Bank, the Asian Development Bank, the African Development Bank Group, and the Asian Infrastructure and Investment Bank—rather than it being monopolized by a single body.

As such, the loss and damage fund is a lifeline that should not be subject to the whims and bureaucracy of distant institutions. Ultimately, we must operate not within existing archaic and politically compromised structures but be bold enough to create entirely new structures capable of meeting the unprecedented needs of vulnerable nations struggling against climate change. This means reforming the Bretton Woods institutions, such as the World Bank and IMF, creating a new global development bank, and establishing an entirely new global financial transaction tax.

But making all this a reality requires the support of industrialized Western nations and a radical departure from the norm. The survival of our planet calls for it. The alternative? Loss and damage might end up in the dustbin of failed climate efforts.[128]

On the first day of COP 28, the loss and damage fund was finally agreed on. The host country, UAE, promised to donate US $100 million, while other donors included Germany (US$100 million), UK (US$75 million), US (US$24.5 million) and Japan (US$10 million).

At the same time, the authoritative economist on inequality, Thomas Piketty, argued that it is not enough to look at the issue of climate justice from a developed *versus* developing nation perspective but we must also focus on the inequality inherent *within* developed nations. He cited the example of Gilets Jaunes in his native France as well as the fact that there is a huge gap in carbon footprints between the ultra-rich and the rest of society. Piketty argues that a progressive carbon tax along with bans on unnecessarily high carbon emitting luxuries are important instruments to address this inequality. The top 1 per cent of society enjoy private jets, large private vehicles, and short-distance flights that leaves a massive carbon footprint but are essentially superfluous.[129]

International Developments

US President Joe Biden signed the IRA into law in 2022. This is the most significant clean energy and climate action legislation passed in the country. US$783 billion will be spent on energy

---

[128] Nik Nazmi Nik Ahmad, 'Western Nations Are on the Cusp of Usurping Loss and Damage Fund. This Could Derail COP28', *Euronews*, November 3, 2023, https://www.euronews.com/2023/11/03/western-nations-are-on-the-cusp-of-usurping-loss-and-damage-fund-this-could-derail-cop28 [Accessed November 20, 2023].

[129] Harvey, Fiona, 'Ban Private Jets to Address Climate Crisis, Says Thomas Piketty', *The Guardian*, November 22, 2023, https://www.theguardian.com/environment/2023/nov/22/ban-private-jets-to-address-climate-crisis-says-thomas-piketty [Accessed March 5, 2024].

and climate change, including on drought resiliency. The IRA incentivizes clean energy and equity-centred environmental investments. It is forecasted that other than reducing inflation (as highlighted in the name of the legislation), it will also minimize the government deficit.[130]

The idea of the IRA is to 'crowd in' private investment to prepare the US for economic shocks and build up the country's energy security. By focusing on climate resilience, it allows for more sustainable and shock-proof economic growth that, in turn, will lead to a healthy, growing economy that will sustain America's finances.

In the EU, CBAM is designed to tackle carbon leakages to address any potential gap with other countries, as the EU becomes more ambitious with its climate agenda. It will put a price on carbon emissions for goods coming into the bloc. This will deal with the prospect of EU companies moving production of carbon-intensive products overseas and substituting EU products with carbon-intensive imports. The CBAM entered its transitional phase in October 2023.

Generally, while the IRA takes more of a 'carrot' approach, the CBAM is arguably the 'stick'. The fear is that other than laudable climate goals, there are also nationalistic objectives implicitly embedded in those policies. Malaysia will work with other developing countries, in particular our fellow ASEAN members, to forge our own policy framework that considers our circumstances and challenges. As a country that trades heavily with the EU and the US, we will also have to ensure that our companies can meet with these new requirements. An ambitious energy transition goal is a competitive advantage to benefit from these international developments. Most importantly, however, it is an existential necessity with the climate challenges that we face.

---

[130] The naming of the legislation is a good example of highlighting to voters the socio-economic advantages of climate and environmental policies.

COP 28 saw markedly different positions from various countries on the question of shifting away from fossil fuels. For decades, Saudi Arabia and other major oil and gas producers had blocked any attempt to include fossil fuels in global climate agreements. It was only at COP 26 in Glasgow that the dirtiest fossil fuel, coal, was included in the agreement. The US and EU were pushing for a phaseout of unabated fossil fuels, allowing room for carbon capture. Small island states were pushing for more radical action to address climate change.

Finally, the phrase that was agreed upon in COP 28 was for a 'transition away' from fossil fuels. It was not as strong as the 'phase out' or 'phase down' wording demanded by some countries and activists, but the fact that fossil fuels have finally been accepted as a problem and carbon capture has been emphasized on less is significant progress. Malaysia's position is that the reduction must be gradual and that the developed world must take the lead rather than forcing all countries to make the transition at the same time.

Show Me the Money!

In February 2023, the IMF estimated that trillions are needed annually across the world for climate finance to achieve our climate goals by 2050. But only around US $630 billion is being spent annually on climate finance at the global level and only a fraction of this goes to developing countries.[131]

There have been announcements by developed countries to assist developing countries with climate change. While the announcements were ambitious, the reality is different because the real amounts channelled fall short of the big pledges

---

[131] Bo Li, 'Scaling up Climate Finance for Emerging Markets and Developing Economies', *International Monetary Fund*, February 27, 2023, https://www.imf.org/en/News/Articles/2023/02/28/sp022823-scaling-up-climate-finance-for-emerging-markets-and-developing-economies [Accessed September 30, 2023].

announced to the world. It is as if we see huge cheques being signed by the developed countries for the Global South yet, often, we are unable to cash these cheques!

During COP 15 in Copenhagen, developed countries committed to channelling US $100 billion *annually* by 2020 to developing countries for urgent climate action. They have already fallen short of that pledge made in Denmark.

In November 2022, there was a lot of excitement when Indonesia launched the JETP with its partners including the US, EU, and Japan. US$20 billion will be utilized to allow the country to pivot from fossil fuels (particularly coal, which is locally produced) to renewable energy. Yet, halfway through, in 2023, the investment plan had to be pushed back as the JETP secretariat said more time was needed to find a credible pathway for transition. There were arguments about whether the money would come as grants or loans. Furthermore, while initially US$20 billion seems like a lot, in reality, an estimated minimum US$1 trillion is needed for Indonesia's transition by 2060.[132]

When we talk about climate justice, we must also look within our boundaries. We must not forget SMEs. I like to remind banks and financiers that SMEs do not have chief sustainability officers to help them navigate new complicated ESG requirements both internationally and domestically. Yet, the Malaysian economy rests on the backs of SMEs that form 97 per cent of Malaysia's local business establishment and employs almost half of the country's workers.[133]

---

[132] Kurnia, Erika, 'Jokowi: Indonesia Needs IDR 15,439 Trillion for Energy Transition', *Kompas*, December 2, 2023, https://www.kompas.id/baca/english/2023/12/02/en-indonesia-butuh-1-triliun-dollar-as-untuk-transisi-energi [Accessed March 5, 2024].

[133] World Bank, 'Malaysian SME Program Efficiency Review', March 2022, https://documents1.worldbank.org/curated/en/099255003152238688/pdf/P17014606709a70f50856d0799328fb7040.pdf [Accessed October 10, 2023].

Malaysia (including Prime Minister Anwar Ibrahim in his earlier days in government) has been playing a major role in the growth of the modern Islamic finance industry. While assets under Islamic finance are still a small fraction of financial assets in general, the development has been spectacular. Islamic finance is underpinned and governed by Islamic law, the sharia. Usury, speculation, gambling, and prohibited products and services deemed harmful to human society—alcohol, pork, pornography, etc.—are strongly prohibited.

Lately, ethical investing in the form of ESG has gained prominence. Investors and consumers are looking beyond short-term profits but are taking a multi-faceted approach instead. In fact, financially, it makes sense for most businesses to protect long-term sustainability. When I was interviewed at the keynote session for *The Financial Times'* Moral Money Summit Asia, I said that Islamic finance should take a more inclusive approach to include modern ESG principles in-line with the higher objects of the sharia, the *maqasid sharia*. This was first outlined by influential eleventh-century theologian and philosopher al-Ghazali as the purpose of divine law: the protection faith, life, intellect, progeny, and property. While some might find this unconventional, contemporary secular ethical finance has religious roots as well: it started with Christian Quakers and Methodists.

Start 'Em Young

Children in two schools near the Kim Kim river in Johor started coughing, feeling sick, and having breathing difficulties in March 2019. Some even fainted. They smelled a strong and pungent odour in their school. It was later discovered that a tanker from an illegal tyre recycling factory had poured nearly 2.5 tonnes of marine oil into the river early on the morning of the incident. This released dangerous methane and benzene fumes. The incident affected nearly 6,000 people and 2,775 people, mostly children, received immediate treatment.

Image 22: I appointed civil society representatives to be on the Climate Change Consultative Panel, including representatives from indigenous groups and youth. This becomes part of the consultation process for the ministry in enacting new policy.

Climate change and environmental crises affect children more adversely, despite them being caused by the sins of their parents. The older generation is responsible for the problems plaguing the planet. Children are the most vulnerable to environmental problems. Children also do not have a say in the democratic process. As a result, many policies and legislations do not address children's rights. Many young activists tell me, in my town hall and consultative sessions with them, to include children as part of the solution. In fact, as mentioned, we should incorporate the principles of long-termism as a crucial pillar of designing our policies today. We have decided to form a youth cluster to include children in our Climate Change Consultative Panel. Through Yayasan Hijau Malaysia, the environmental foundation arm of NRES, we have increased engagement events with schools, universities, and local communities.

Children under five years are considered extremely vulnerable to diseases caused by climate change and pollution. Diseases such as malaria and malnutrition are predicted to get worse due to climate change. Children are also more likely to be victims of heatwaves as well as flooding. Previously, boys and girls exposed to leaded fuels were found to have lower IQ.[134] At the same time, psychologically, children are more vulnerable to any crisis or trauma.[135]

But there are opportunities as well. It is easier to develop positive habits among the younger generation. By treating them with respect, not condescension, we can create better awareness about the need to protect the environment and develop climate resilience. Parents can start with small, basic actions: separating the trash, composting, and yes, explaining why wasting electricity and water is not only bad for the family wallet but also for the planet. Be creative and make it fun, by all means. What ultimately matters is inculcating positive values among the younger generation.

---

[134] Heidari, Serve, Shayan Mostafaei, Nazanin Razazian, et al., 'The Effect of Lead Exposure on IQ Test Scores in Children Under 12 Years: A Systematic Review and Meta-Analysis of Case-Control Studies', *Systematic Reviews*, vol. 11,1. 106. May 30, 2022, https://www.ncbi.nlm.nih.gov/pmc/articles/PMC9150353/ [Accessed March 15, 2024].

[135] Mann, Michael E., Raymond S. Bradley, Malcolm K. Hughes, 'Northern Hemisphere Temperatures during the Past Millennium: Inferences, Uncertainties, and Limitations', *Geophysical Research Letters*, vol. 26, 6. March 15, 1999, pp. 759–762, https://agupubs.onlinelibrary.wiley.com/doi/epdf/10.1029/1999GL900070 [Accessed March 5, 2024].

Image 23: Meeting eleven-year-old Braden Yong who spoke at COP28 about the climate challenges faced by Malaysia and the world

At the same time, as any parent can attest, having a child entirely changes our perspective on life. We become less selfish and think ahead about our children's future. I spoke to NGOs that work with children and see the impact of how raising awareness about climate change and the environment among the young eventually changes the habits and behaviours of parents as well. A clear example is the youth climate change activism phenomenon Greta Thunberg. She became deeply concerned about climate change as a schoolgirl and started a climate strike to highlight the issue when she was still a teenager. In fact, she influenced her parents to care and do more about climate change!

Ou Hongyi's story is probably even more remarkable. Based in Guilin, China, she is the same age as Greta and stopped going to school after watching Al Gore's *Inconvenient Truth*. Hongyi was inspired by Greta to protest alone for six days in front of the local authority building. Then, she was questioned by the police. The authorities blocked her WeChat account. Her parents found it a challenge to accept all her choices, including postponing going to university, but adopted some of her choices, such as becoming vegetarian.

Decades before Greta and Hongyi, there was Severn Cullis-Suzuki. She established the ECO as a nine-year-old. During the 1992 Earth Summit, Severn, who was twelve then, attended the conference in Brazil with other members of ECO. She addressed the plenary session of the summit, and a recording of the video went viral across the world when it was uploaded by a Malaysian on YouTube in 2008. The video to date has 32 million views. Severn became known as 'the girl who silenced the world for five minutes'. She explained in her speech why she pooled money to fly almost 10,000 kilometres from Canada to Brazil. Some of the politicians and diplomats in attendance looked very uncomfortable to be told hard truths directly by a young girl.

> I am fighting for my future. Losing my future is not like losing an election or a few points on the stock market. I am here to speak for all generations to come [. . .]
>
> I am afraid to go out in the sun now because of the holes in the ozone. I am afraid to breathe the air because I don't know what chemicals are in it [. . .]
>
> In my country, we make so much waste, we buy and throw away, buy, and throw away, and yet northern countries will not share with the needy. Even when we have more than enough, we are afraid to lose some of our wealth, afraid to share [. . .]
>
> Do not forget why you're attending these conferences, who you're doing this for—we are your own children. You are deciding what kind of world we will grow up in. Parents should be able to comfort their children by saying 'everything's going to be alright', 'we're doing the best we can' and 'it's not the end of the world'.

But I don't think you can say that to us anymore. Are we even on your list of priorities? My father always says, 'You are what you do, not what you say.'

Well, what you do makes me cry at night. You grown-ups say you love us. I challenge you, please make your actions reflect your words. Thank you for listening.[136]

---

[136] Lee Kim Seong, 'The Girl Who Silenced the World for 5 Minutes', *YouTube*, April 18, 2008, https://www.youtube.com/watch?v=TQmz6Rbpnu0 [Accessed December 29, 2023].

Conclusion

> Our children,
> will never enjoy the beach,
> playing with many friends together.
> Nature is now old—its age is more than million,
> we will work hard to prevent catastrophe.
> The ozone layer is getting thinner day by day,
> it's getting eaten by hazardous fumes.
> This is the work of man,
> all the things in the world,
> cannot last,
> for eternity.
>
> —Translation of 'Hijau' (Green),
> a popular song by Zainal Abidin

Since the 2022 elections that led to the formation of the post-coalition Unity Government, there has been a much-talked-about 'green wave', where the Islamic party, PAS, and a Malay nationalist party, the Malaysian United Indigenous Party (PPBM) made gains in Malay areas. Green is the colour of PAS. As I have tried to argue, the challenges that we face to protect our environment and build climate resilience are not scientific or economic. Instead, the challenge is political, so we need a new green wave, one that centres on a clear environmental agenda to save the planet, rather than the divisive identity politics of race and religion.

Throughout this book, I have sought to outline the different environmental and climate challenges Malaysia faces. I have tried to make it as accessible as possible, to a wide spectrum of readers, to cut through the technical jargon and complicated science. I have also highlighted the delicate balance between idealism and pragmatism that politicians often navigate to understand our great existential mission. What readers who care about the future should take away is that we must build broad alliances to progress with ambitious goals. We need to cherish the small chunks of doable reforms that we can achieve and keep up the momentum. This is easier said than done, but without trying, we are giving up too easily. We do not have the choice of giving up.

The Covid-19 pandemic majorly impacted the planet, including the environment and climate. As mentioned, while this was foreseen, because of increased human-wildlife conflict and shrinking habitats, no one could accurately predict when and where it would happen. Climate change, on the other hand, has been predicted for many years and has been spoken of widely since the 1990s. The timeline and impact of rising temperatures on the planet have been clearly elucidated. The hockey stick graph clearly shows that while the world initially enjoyed a long-term, mild cooling effect between the last 500 and 2,000 years, there has been a rapid warming since the twentieth century, following the Industrial Revolution.[137]

Among my first memories of environmental issues is from the early 1990s. I was in primary school and followed current issues and news from the world of science closely even as a kid. The first Malaysian election that I followed, rooting for the

---

[137] Mann, Michael E., Raymond S. Bradley, Malcolm K. Hughes, 'Northern Hemisphere Temperatures during the Past Millennium: Inferences, Uncertainties, and Limitations', *Geophysical Research Letters*, vol. 26, 6. March 15, 1999, pp. 759–762, https://agupubs.onlinelibrary.wiley.com/doi/epdf/10.1029/1999GL900070 [Accessed March 5, 2024].

opposition, was the 1990 General Election. Meanwhile, there was a superhero cartoon series called *Captain Planet and Planeteers* on television. The Planeteers were Kwame from Africa, Wheeler from North America, Linka from the Soviet Union, Ma-Ti from South America, and Gi from Asia.

Little did I know then that these five were based on real, young environmental activists of that time. In preparation for the 1992 Earth Summit in Rio de Janeiro, Chee Yoke Ling, a Malaysian activist with the Sahabat Alam Malaysia, and four other young environmentalists met with *Captain Planet*'s co-producer, Barbara Pyle. Thus, the five of them became real-life inspirations for the Planeteers. Gi was based on Chee (Barbara pronounced 'Chee' as 'Gi'). When the Planeteers were unable find solutions on their own, they summoned Captain Planet, who combined all of their enhanced powers, along with a few additional (typical) superhero abilities such as flight and invincibility. Throughout this book, we have seen how the Earth Summit in Rio established the framework for climate diplomacy, along the lines of the Bretton Woods system for the financial world.

This was around the time that Zainal Abidin's song 'Hijau' became popular in Malaysia. For someone like me with Kelantanese roots who grew up speaking the dialect at home, the Kelantanese lines in the song were catchy. But reflecting on Captain Planet and the global efforts with regards to the ozone layer as well as removing lead from petrol offered us some concrete lessons for fighting climate change—challenging as the task may seem.

The ozone layer is a belt with high concentration of ozone in the earth's stratosphere. It protects life from the sun's harmful UV rays. In the 1970s and 1980s, it was discovered that the layer had been depleting—the much-talked about ozone hole. Much of this was caused by CFCs, which are used for refrigeration and in aerosols. When this was established, different countries began to act, which culminated in the Montreal Protocol that drastically reduced the use of CFCs. Today, the depletion in the ozone layer is slowing down or has even ceased. The ozone layer is expected to return to 1980 levels by the middle of the twenty-first century.

Lead is called *plumbum* in Malay, from the Latin word for the material. It was extensively used during the Roman Empire for piping; hence the origin of the English word 'plumbing'. There has been speculation that lead poisoning contributed to the decline of the Roman Empire due to its effects mentioned earlier, such as lower IQ and even possibly higher infertility. Now, the situation is far worse due to the use of leaded petrol.

In July 2021, Algeria stopped the sale of leaded petrol to cars, becoming the last country to do so. The first was Japan, forty-five years ago. Malaysia banned leaded petrol in 2000. Two years later, a global campaign began to stop the sale of leaded fuel after scientists discovered that it caused over one million premature deaths. It was found to cause cancers, strokes, and coronary diseases. It also affects the brain. A study by paediatrician and child psychiatrist Herbert Needleman revealed that children exposed to lead have an IQ up to seven points lower than other children! Their attention spans are likely to be low and they may also have behavioural problems. The authors of *Freakonomics* attributed the dramatic decline in crime rates in the developed world of the 1990s to diverse causes (such as increased legalized abortion, changing alcohol consumption patterns, and reduced distribution of crack cocaine, among others). Interestingly, this was also the time when unleaded petrol replaced leaded petrol. A few studies attributed the reduction in crime to this.[138]

---

[138] Reyes, Jessica Wolpaw, 'Environmental Policy as Social Policy? The Impact of Childhood Lead Exposure on Crime', *The BE Journal of Economic Analysis & Policy*, vol. 7, 1. 2007, https://jwreyes.people.amherst.edu/papers/LeadCrimeBEJEAP.pdf [Accessed September 13, 2023];
'Phase-out of Leaded Petrol Brings Huge Health and Cost Benefits – UN–Backed Study', UN News, October 27, 2011, https://news.un.org/en/story/2011/10/393292-phase-out-leaded-petrol-brings-huge-health-and-cost-benefits-un-backed-study [Accessed September 13, 2023].

These incidents showed what the world can achieve by coming together and adhering to science. But science must have a *purpose*. It must be guided by ethics. Remember, much of the damage that we have done to Mother Nature has come from how we have *utilized* science and technology to satisfy our uncontrolled greed and thirst for resources since the Industrial Revolution. At the end of the day, science is merely a *tool*, like a knife. It can be used for good or bad. The arrogance of humans, who think we have the right to shape the world in own our image, has now led nature to seek revenge on us. Environmentalist Gus Speth's quote, which I referred to at the beginning of the book, about the planetary crisis fundamentally being a spiritual problem comes to mind.

Recent times have also shown how central a stable, functioning government is to build climate resilience and protect a country's environment. The University of Notre Dame ranked countries most vulnerable to climate change in 2021. Fourteen of the twenty-five worst performing countries are those that are in conflicts, such as the Democratic Republic of Congo, Yemen, and Afghanistan.

Like many countries in the Levant, Palestine suffers from severe water stress. In general, Israelis use four times more water than Palestinians. But too much rain also causes floods. For those living in makeshift shelters because of the ongoing conflict, the flooding and cold weather during winter brings widespread suffering. Gaza is one of the world's most densely populated areas. The Israeli occupation and blockade prevent Gazans from doing what most other communities would do—migrate to more fertile lands. During COP 28, leaders from across the world—including South Africa, Türkiye, Jordan, and Colombia—raised the issue of Palestine.

In an interview with Saudi-based *Arab News*, I mentioned that COP 28 must address the conflict in Gaza from a climate and environmental perspective as well. It has become difficult to breathe the air there while the water is now difficult to drink because of the bombings by the Israeli military. B'Tselem, an Israel-based human rights organization found that because of Israel's control over water reserves in Palestine, an Israeli can now use *five* times more water than a Palestinian in Gaza or the West Bank. Access to sanitation is woefully inadequate in the region. Palestinians also are losing valuable land due to sea-level rise.

'The occupation robs Palestinians of their right and ability to improve climate resilience due to the inordinate control over Palestinian land, water, and other vital natural resources,' I told *Arab News* in an interview.[139]

The seeming indifference to the fate of the Palestinians stands in contrast to the West urging the world to support them during Russia's invasion of Ukraine. There, much attention was given to Russia's ecocide. The hypocrisy has not been lost on the world. Fiona Hill, a former senior official at the US National Security Council, said:

> Countries in the Global South's resistance to US and European appeals for solidarity on Ukraine are an open rebellion. This is a mutiny against what they see as the collective West dominating the international discourse and foisting its problems on everyone else, while brushing aside their priorities on climate change compensation, economic development, and debt relief. The Rest feel constantly marginalized in world affairs.[140]

For so long, conventional wisdom was that the goal of economic policies was to promote unlimited growth. Getting higher GDP

---

[139] Laskowska, Natalia, 'Malaysia Says COP 28 Should Address "Devastating" Climate Toll of Israel's Onslaught on Gaza', *Arab News*, December 2, 2023, https://www.arabnews.com/node/2419071/world [Accessed December 22, 2023].

[140] Hill, Fiona, 'Lennart Meri Lecture 2023: Ukraine in the New World Disorder', *International Centre for Defence and Security*, May 13, 2023, https://lmc.icds.ee/lennart-meri-lecture-by-fiona-hill/ [Accessed December 27, 2023].

growth was a bit like getting the highest mark in an exam. The belief was that this was not only possible but also a necessity to compete with other countries. GDP measures the value of finished products and services in a country at a particular point in time. In 1972, the Club of Rome released a report, *The Limits to Growth*, which looks at the impact of economic and population growth on the world's limited resources. At that time, the report received widespread criticism for being too pessimistic and not considering the impact of technological improvement. But twenty to thirty years later, more attention has been given to the work as the impact on the environment, climate, and biodiversity becomes more apparent and the idea is given more consideration by the scholars, politicians, and scientists.

In fact, the Big Oil and gas players were aware of the damage done by fossil fuels as far back as the 1950s! The Keeling Curve is a graph depicting the daily record of carbon dioxide concentration taken in Hawaii from 1958 until today. Dubbed as a critical scientific discovery, it is the first proof of a significant, continuous increase of carbon dioxide in the earth's atmosphere. It is named after scientist Charles David Keeling, who started the work and monitored it until his death in 2005. Recently, it was discovered that the oil and gas industry had knowledge of increasing carbon dioxide levels in the atmosphere due to fossil fuels and their impact on the planet as early as four years before the work on the Keeling Curve started. A research proposal was sent to the Air Pollution Foundation—which was funded by, among others, the American Petroleum Institute and the Western Oil and Gas Association—to look at the impact of fossil fuels on the climate. The foundation was just formed at that time to look at the Los Angeles smog problem, and it approved a grant for the research.[141]

---

[141] John, Rebecca, 'New Evidence Reveals Fossil Fuel Industry Sponsored Climate Science in 1954', *DeSmog*, January 30, 2024, https://www.desmog.com/2024/01/30/fossil-fuel-industry-sponsored-climate-science-1954-keeling-api-wspa/ [Accessed January 31, 2024].

In July 1977, a scientist briefed his company's top executives that carbon dioxide released from fossil fuels was probably warming the earth, which is harmful to human beings and the environment. Concerned about the small window to address climate change, he recapped the presentation in an internal memo. He estimated that carbon dioxide levels would double in the atmosphere by 2050. This would lead to an increase in global temperatures of 2 to 3 degrees Celsius, on point with what is predicted today.[142] The same company was manufacturing solar panels based on the concern that oil was running out (and to combat the oil price hikes of the 1970s).[143]

The company being discussed is Exxon, which would later buy Mobil, another oil and gas company, and is today known as ExxonMobil. Not only was Exxon's scientist James F. Black's prediction correct but the company's forecasts on climate change were also highly accurate until the 1980s, in-line with academic as well as government experts. Ten years after Black's memo, however, they became one of the prime movers in climate change denial, forming the Orwellian Global Climate Coalition in 1989. This organization of big corporations with vested interests played a key role in the decision for the US not to be part of the Kyoto Protocol.

In contrast, most of the developing countries do not deny climate change and environmental degradation. We recognize the fact that the old way of doing things, of growth at all costs, is not sustainable. We know that unbridled capitalism is not the path we

---

[142] Black, James F, 'The Greenhouse Effect', Internal Document, 1978, https://ia801806.us.archive.org/20/items/aQwayback/exxon/James%20Black%201977%20Presentation.pdf [Accessed December 29, 2023].

[143] Hsu, Andrea, 'How Big Oil of the Past Helped Launch the Solar Industry of Today', NPR, https://www.npr.org/2019/09/30/763844598/how-big-oil-of-the-past-helped-launch-the-solar-industry-of-today. [Accessed July 30, 2024].

want to pursue because that is the road the developed world took and look where it has brought us today. Yet, the need to uplift the standard of living of our people, to bring them out from poverty and to secure dignified lives for them is central to our task. That is why it is important for us to look at the doughnut economics model referred to earlier in designing our policies and not stick to beggar-thy-neighbour strategies.

As mentioned at the beginning of the book, prior to my appointment as NRECC minister, my interventions about environment and climate change were rather limited. But, in one such instance, in 2019, as a government backbencher, I wrote a piece on climate change entitled 'The global "Climate Strike" and haze must be a wake-up call':

> Sustainability is inextricably linked with economic progress: we cannot have one without the other.
>
> There have been many suggestions of how ordinary Malaysians can do their part to reduce climate change, including everything from using less single-use plastic to taking public transportation more.
>
> These are all worthy ideas that we ought to adopt if we can.
>
> But ultimately, what is needed is for governments to lead the way by their policies and actions.
>
> Malaysian leaders and their counterparts elsewhere cannot demand that their people change their lifestyles without first taking a good, hard look at their own.[144]

---

[144] Nik Nazmi Nik Ahmad, 'The Global "Climate Strike" and Haze Must Be a Wake-Up Call', *The Star*, September 28, 2019, https://www.thestar.com.my/opinion/letters/2019/09/28/the-global-039climate-strike039-and-haze-must-be-a-wake-up-call [Accessed June 8, 2023].

One of the definitive speeches of the twentieth century was the 'moonshot' speech by President John F. Kennedy at Rice University in 1962, studied not just by space hobbyists and politicos but also as a sterling example of how to win over public opinion and motivate a nation:

> We choose to go to the Moon in this decade and do the other things, not because they are easy, but because they are hard; because that goal will serve to organize and measure the best of our energies and skills, because that challenge is one that we are willing to accept, one we are unwilling to postpone, and one we intend to win, and the others, too.[145]

The legacy was significant even for my generation, although I was born two decades later, long after JFK's passing. Americans were pessimistic about winning the space race against their rival, the Soviet Union. The latter had put the first satellite and then the first man in space (and between that, Laika the dog, the first animal to go to space). The cost of landing a man on the moon by 1970 was over US $150 billion in today's money and was obviously controversial among the American taxpayers. But that visionary dream was not only about that physical mission; the sheer effort galvanized the resources and research that spawned new technologies. It was also, despite the ongoing Cold War, a call for international cooperation for space exploration. I firmly believe that addressing climate change will be the 'Moonshot' challenge of our generation.

In 2019, the former US Secretary of State and future Climate Envoy John Kerry, along with the Congressman representing

---

[145] Kennedy, John F., 'Address at Rice University on the Nation's Space Effort, September 12, 1962', *John F. Kennedy Presidential Library and Museum*, https://www.jfklibrary.org/archives/other-resources/john-f-kennedy-speeches/rice-university-19620912 [Accessed July 20, 2024].

Silicon Valley—Ro Khanna called for the US to not lose to China in the 'Green Race' with regards to clean energy. China has an edge with regards to producing, selling, and installing wind turbines, solar panels, batteries as well as electric vehicles. It is also leading in terms of intellectual property related to renewable energy. The fastest train from Chicago to New York takes nineteen hours whereas if they had the same high-speed trains as China, it would be only a fifth of that.[146]

A year later, the-then Duke of Cambridge Prince William launched the Earthshot Prize in collaboration with the Aga Khan Development Network—with less geopolitics involved. Inspired by Kennedy's Moonshot, the Earthshot is an attempt to replicate the stirring motivational vision to address the biggest environmental and climate challenges of our generation. There are five Earthshots: protect and restore nature, clean our air, revive our oceans, build a waste-free world, and fix our climate. As stated by Prince William:

> The global response to the Covid-19 pandemic is evidence that all this is possible. The funds flowing into the recovery effort demonstrate how much can be achieved when those in positions of power come together and decide to act. We've built hospitals overnight, repurposed factories and poured billions into the search for a vaccine and better treatments. And we've been inspired by heroes emerging in every community across the world.[147]

---

[146] Kerry, John and Ro Khanna, 'Don't Let China Win the Green Race', *The New York Times*, December 9, 2019, https://www.nytimes.com/2019/12/09/opinion/china-renewable-energy.html [Accessed January 14, 2024].

[147] Butfield, Colin and Jonnie Hughes, *Earthshot: How to Save the Planet*. London: John Murray, 2021, p. xiii.

Image 24: Greeting William, Prince of Wales, at the Earthshot Prize 2023. Talking to him is Vjosa Osmani, president of Kosovo. Left to right: Vivian Balakrishnan, Singapore's minister of foreign affairs; Grace Fu, Singapore's minister of environment and sustainability; and Ho Ching, chairman of Temasek Trust and spouse of the prime minister of Singapore.

I was invited to attend the 2023 Earthshot Prize awards ceremony in Singapore. While my son was envious that I managed to watch OneRepublic and Bastille live as they provided the entertainment for the day, I was truly inspired by the various finalists from across the world who came with innovative ideas to address the five Earthshots. Acción Andina based in the Andes was inspired by the ancient Inca notions of 'Ayni and Minka', working together for the common good. They work with the local population to replant native species on the mountain that had been ravaged by deforestation and mining, as well as depleting glaciers that disturb water supply and fertility of farmland. Another initiative, Boomitra, works with farmers that improve their farmland to increase its carbon storage, and then matches them with companies looking for carbon credits.

Scientists have confirmed that the world has just experienced its hottest summer in history in 2023. We have heard the Secretary-General gravely declare that 'Climate breakdown has begun.' Even Malaysia is seeing an increase in the adverse impacts of climate change, with increasing temperature, rising sea levels, intensified monsoons, and erratic weather patterns disrupting livelihoods and degrading local ecosystems. As such, we have not a moment to lose. I believe Malaysia and Malaysians are awakening to the challenge.

In Anwar Ibrahim's maiden country statement as the prime minister to the UN General Assembly in September 2023, he focused on several key issues including climate change:

> Malaysia is doing its part by developing low-carbon and renewable energy roadmaps to implement mitigating and adaptation strategies. The newly launched National Energy Transition Roadmap should aid us in achieving our NDCs, as well as lighting the path towards our net-zero aspirations.
>
> Discussions on climate ambition in the absence of equity, justice, and the necessary means to assist and empower countries to undertake greater climate action is an exercise in futility.
>
> We also urge the developed countries to fulfil their commitment of mobilising US $100 billion a year to support climate ambition endeavours of developing countries while recognising that trillions of dollars per annum will be needed in the near future.

As NRECC minister, I wanted to make environment and climate change a priority for the government. I have succeeded due to the strong support of the Prime Minister. Under the Madani Agenda, sustainability is among the key pillars. As part of the Madani Economic Framework, energy transition, climate resilience, and universal supply of electricity and water are all focus areas.

On the first day the new, post-2022 Parliament sat, I wrote to the Speaker pushing for the establishment of a Select Committee to oversee my portfolio. As a result, a Special Select Committee on the Environment, Science, and Plantations has been established.

Malaysia had a high-profile presence at COP 28 in Dubai. The king, Sultan Abdullah, and his son, the Regent of Pahang Tengku Hassanal Ibrahim Alam Shah, were present along with a few ministers. Our pavilion hosted events by government departments, the corporate sector, and civil society, not just from Malaysia but also from various other countries. In Malaysia's country statement, I emphasized our commitment to fight climate change. We applauded the setting up of the loss and damage fund and communicated our willingness to be even more ambitious, if there is more support from the developed world.

> The goalposts should not keep changing. It should not be the case that some wealthy countries, including those who have benefitted from exploiting the environments of other nations in the past, are now allowed to be 'pragmatic' when it comes to climate change, while developing ones are held to stricter and inflexible standards or subject to unilateral trade measures.

Far away from the gilded conference halls of Dubai, I have had the chance not only to listen to scientists and economists but also to ordinary Malaysians impacted by climate change—from the rich to the poor; urban or rural; from the Peninsula, Sabah, or Sarawak. I visited Sabak beach, not far from where the Imperial Japanese Army landed during the Second World War in Kelantan, which had been plagued by tidal floods. I was told that more than thirty plots of land, complete with titles, were now part of the seabed. The residents begged for help, as the floods had started reaching the local school: 'The government has spent so much to build the school. Please ensure we don't eventually lose the school to the sea.'

I also went to the Tok Jembal beach in Kuala Nerus, Terengganu. The coastal road was closed, as the sea reached the road boundary, swallowing the beach in its wake because of severe coastal erosion. The famous stalls selling *cholek* or deep-fried seafood had to close. From devastating floods and children dying in heatwaves, climate change is happening right in front of us.

In the introduction, I mentioned two green controversies that I had to face as a legislator—once at the state level and once as an MP. Both involved protests against developments from the local community. As politicians, we often deal with so-called nimbyism. That is fair: who would want to live next to a crowded development when the area is zoned for recreation? Nor would anyone want to stay next to a high-rise condominium on a steep slope that had experienced landslides after landslides all these years. But while climate and environmental problems affecting the planet may seem distant, whatever affects our planet's fragile ecosystem will have a long-term impact that will affect us all. Ultimately, the entire planet is our backyard.

Renowned climatologist Michael E. Mann remarked that this is 'the new climate war', where polluting corporate giants—following the playbook of tobacco and gun lobbies—pursue a path of deflecting responsibility for the climate crisis from themselves to individuals. The story of Exxon is a good example of this strategy. We are told to have less children. Some vegan activists, particularly from the developed world, like to guilt-trip those eating meat or consuming animal products, when impoverished communities, especially in the developing world, lack sufficient protein. Yes, eating less meat or avoiding flying does help, yet we need more than individual actions. In fact, Mann argues that a purely individualized approach tends to allow corporates to get away with their damaging ways as we pat ourselves on our back for what we have done. Just this being our response will not move the needle whereas government policies and corporate actions will get things moving at a different scale altogether.[148]

We must also be wary of greenwashing by government and corporations, of making climate and environmental action merely a public relations tool without any real substance driving it. Often, in cases of greenwashing, the money spent on advertising to the public that a certain product or service is 'green' is more than what is used for taking actual green action. On the other hand, there are those who might not care enough or think that climate change and pollution are not serious problems. Worse, some might develop a sense of fatalism: pointing to how little difference we can make with individual actions—they argue we might as well stop bothering and just live in the here and now.

Cost is often cited as a constraint that limits how much we can act on climate change. It's true that the process is not cheap. But the cost of doing nothing is even worse: damage from floods, forest fires, disappearing coastlines, and even deaths caused by polluted air and rivers.

Investment in climate resilient infrastructure pays off. In the Netherlands, the Delta Programme was introduced in 2007 with the purpose of flood mitigation, water supply, and spatial planning to take a more contemporary approach regarding the age-old Dutch challenge of protecting their low-lying country. It brings together the government, private sector, academia, and the community. Nature-based solutions, including adapting to the river, are prioritized. In early 2023, Türkiye and Syria were hit by a massive earthquake, one of the largest recorded ever in West Asia. It was followed by 30,000 aftershocks over the next three months, with nearly 60,000 deaths and nearly US$120 billion in damages. One of the positive stories that emerged was that twenty-four schools built in the affected areas under the

---

[148] Mann, Michael E., *The New Climate War: The Fight to Take Back Our Planet*. London: Scribe Publications, 2022, p. 3.

Education Infrastructure for Resilience Project together with the cooperation of the World Bank and the EU, among others, survived the disaster.

Governments can make the necessary systemic changes. The onus is on politicians to take leadership and convince the public. We must have the courage to lead from the front, not follow meekly from behind. We must provide hope to the people. It must and can be done.

I began this book with a quote from Sadiq Khan's book *Breathe* about how the climate change dilemma, in *the minds* of politicians, seems like something that cannot be solved, does not matter in the short term, and is not the biggest concern for voters in general. In short, the environment and climate change are not vote winners. In the 2023 by-election in Boris Johnson's former seat of Uxbridge and South Ruislip, Khan's Labour Party were the favourites to win. The Conservatives had been in charge for thirteen years. The instability of the UK government following Johnson's reign meant that the turnover of the UK's prime ministers was even faster than Malaysia's during that period.

Yet, the Conservatives won the by-election with a narrow margin after turning it into a referendum of Khan's expansion of the ULEZ policy where the oldest, most polluting cars are charged a daily surcharge to enter the zone. The suburban London seat is popular for those who commute to the city, many relying on private cars. In fact—if you remember, the policy was started in the city centre by former Mayor (surprise, surprise) Boris Johnson! But now, the Labour Party leader Sir Keir Starmer pushed Khan to rethink the policy. The fact is that Labour chose not to shout from the rooftops that they were trying to save lives—Khan was reportedly not welcomed by his own party to campaign in the by-election.[149] Thus, to pin the defeat entirely on ULEZ misses

---

[149] Rawnsley, Andrew, 'Labour's Stumble in Uxbridge Shows Starmer How to Put Party on a Firmer Footing', *The Observer*, July 23, 2023, https://www.

the point. In fact, there was an increase of six percentage points for the Labour Party in a constituency that has been won by the Conservatives in every general election since 1970!

The big picture is clear. The planetary crisis is here. Floods are getting worse and the heatwaves even hotter. Forests are disappearing, wildlife is going extinct, and people are struggling to get clean water. All this has set off a chain reaction that will lead to a vicious cycle if we choose to not to do something meaningful and radical.

It is not all doom and gloom, however. In terms of energy, history has shown that technology has had a positive impact on energy efficiency by bringing down the cost of renewables. Prior to the Paris Agreement in 2014, the trajectory in the increase of the planet's temperature was 4 degrees Celsius. Today, thanks to the adoption of clean energy, the increase is projected to be 3 degrees Celsius.[150] We need to work on a reduction of another 1.5 degrees, but we also ought to look at it as a glass being half-full. We managed to reduce it by an entire degree, which is no mean feat. India, the world's third biggest carbon emitter after China and the US, reduced its emissions by a third in fourteen years in 2023. This faster-than-expected reduction was due to a drop in emissions and an increase in forest cover.[151] As mentioned earlier, while the world continues to experience

---

theguardian.com/commentisfree/2023/jul/23/labour-stumble-uxbridge-shows-keir-starmer-how-to-put-party-on-firmer-footing [Accessed July 24, 2023].

[150] Plumer, Brad and Nadja Popovich, 'Yes, There Has Been Progress on Climate. No, It's Not Nearly Enough', *The New York Times*, October 25, 2021, https://www.nytimes.com/interactive/2021/10/25/climate/world-climate-pledges-cop26.html [Accessed October 3, 2023].

[151] Singh, Sarita Chaganti, 'India Succeeds in Reducing Emissions Rate by 33% over 14 Years – Sources', *Reuters*, August 9, 2023, https://www.reuters.com/world/india/india-succeeds-reducing-emissions-rate-by-33-over-14-years-sources-2023-08-09/ [Accessed October 9, 2023].

increased primary forest loss in 2022, both Indonesia and Malaysia registered near-record low levels.

An increasing number of experts are expecting 2023 to be the peak year in annual global emissions, before they go on a downward trend. True, it would have been better if the peak had happened earlier. And yes, now, the decline needs to take place faster and more rapidly, which will not be an easy or painless process. There will be many difficult decisions that we must continue to make. But this is a dose of good news in our net-zero journey. This has been driven by impressive growth in solar and wind -generated electricity. At the same time, people are buying more electric vehicles to replace internal combustion engines, which use petroleum.[152]

I have seen, first-hand, all sorts of innovation and creativity within Malaysia contributing to efforts to protect the environment and build climate resilience. By combining passion and courage, many difficult challenges can be overcome.

On that point: we are at the heart of Southeast Asia, which matters a lot in the world's effort to fight climate change. ASEAN is only 3 per cent of the Earth's surface, yet it is the world's fourth-largest energy consumer, of which 83 per cent comes from fossil fuels.[153] The population of ASEAN in 2023 is estimated to be almost 690 million, and ASEAN's economy is forecasted to be the fourth largest globally in 2030. As the economy of the region grows and millions of people come out of poverty, so too will the demand for electricity rise. At the same time, ASEAN's role

---

[152] Ambrose, Jillian, 'Climate Scientists Hail 2023 As "Beginning of the End" for Fossil Fuel Era', *The Guardian*, December 30, 2023, https://www.theguardian.com/environment/2023/dec/30/climate-scientists-hail-2023-as-beginning-of-the-end-for-fossil-fuel-era [Accessed December 31, 2023].

[153] Bocca, Roberto and Harsh Vijay Singh, 'Why Southeast Asia Will Be Critical to the Energy Transition', *World Economic Forum*, January 16, 2023, https://www.weforum.org/agenda/2023/01/why-southeast-asia-critical-energy-transition/ [Accessed August 14, 2023].

as a carbon sink cannot be ignored. Protected areas in the region cover more than 800,000 square kilometres and of the 24,889 species assessed, 37 per cent are endemic.[154] Indeed, Southeast Asia is home to 18 per cent of all species assessed globally. The region is home to a fifth of the tropical rainforests worldwide. If Malaysia acts alone, the task is challenging. But moving in concert with other ASEAN member states not only allows us a place at the table with the superpowers and developed countries, as well as international institutions, but also gives us more bargaining power.

Sometimes, it seems that in the journey to protect our environment and build climate resilience, we are in a super-competitive footrace with other countries or regions. But this sends the wrong message, as it should be seen as more like a group hike. We need to succeed together if the planet is to have a fighting chance. After all, we have seen countless examples of how communities come together in facing climate disasters and environmental calamities. A Mayan woman from Belize, Florentina Choco, flew to the other side of the world to be trained as a solar engineer by the Barefoot College in India. What was more impressive was that she did not grow up in a comfortable family and had no formal education. In turn, she and two other women, Miriam Choc and Cristina Choc, installed solar panels in villages and schools that benefited 1,000 residents in Belize.

I hope regulators, academics, activists, and consumers—particularly from the developed world—will be able to understand the realities on the ground as we seek to achieve the same vision of fighting climate change and leaving behind a planet worth living in for our children.

---

[154] Lim, Theresa Mundita, Wiraditma Prananta, and Dwight Jason Ronan, 'Building a Community with a Shared Future: ASEAN's Stake in the Post-2020 Global Biodiversity Framework', *The ASEAN Magazine*, April 12, 2022, https://theaseanmagazine.asean.org/article/building-a-community-with-a-shared-future-aseans-stake-in-the-post-2020-global-biodiversity-framework/ [Accessed August 14, 2023].

The Native American and First Nation Iroquois Confederacy had an oral constitution, known as the *Gayanashagowa*. It is thought to date back to the twelfth century. The constitution uniquely underlined the importance of not only the current but also future generations:

> Look and listen for the welfare of the whole people and have always in view not only the present but also the coming generations, even those whose faces are yet beneath the surface of the ground – the unborn of the future Nation.[155]

Imagine that our *only* home is on fire. For so long, each generation inherited the house, took care of it well and then left it to the next generation. We do not have anywhere else to go. And we know we *can* still save the house for ourselves and our children. We know we will soon reach a point where our only home will not be able to be saved, which means *we* will not be saved as well— just as scientists have been predicting what will happen because of climate change for many decades. This is not something like Covid-19 that blindsided us (despite some forecasts of pandemics having existed previously). We are heading into the burning building with our eyes wide open. Continuing to damage the environment means that we add fuel to the fire that burns our home and threatens our lives. Not doing anything means we basically sitting idly—paralysed by apathy or anxiety—as the opportunity for us to escape becomes smaller and smaller. This is the very definition of insanity.

As the damage we see today to the climate, environment, and biodiversity took centuries to reach this level, the work that needs to be done will take generations. But we must start now.

---

[155] Murphy, Gerald, 'The Constitution of the Iroquois Confederacy', Internet Modern History Sourcebook, Fordham University, https://sourcebooks.fordham.edu/mod/iroquois.asp [Accessed September 30, 2023].

If we all do our bit—if governments push for big changes while businesses take a longer-term horizon and if the public at large do their part—there is hope for us, our children, and their children. We must seize the moment to save this burning house of ours for the survival of our species on this planet, for the sake of the next generation. And that hope should be the starting point for action. Because unlike the analogy of the burning home, I repeat, we have nowhere else to go.

It was mentioned how Covid-19 was an example of not only what countries, communities, and corporations can achieve by working together but also how the planet could revive itself. The air and waters were cleaner while wildlife took over cities. But without acting now, we will soon pass the point where the planet will be able to do that, and we will be consigning our children to a world that is beyond repair.

Index

Abdul Hakim Murad, 11
Abdul Khalid Ibrahim, 71
Abdul Razak Hussein, 108, 181
Abdullah Ahmad Badawi, 109
Accra, 69
Afghanistan, 231
African Development Bank Group, 215
African Union, 15, 187
agriculture, 90, 122, 151, 157, 177, 201, 202, 206
Air Kelantan, 65, 68, 69
air pollution, 4, 93, 94, 96, 98, 101, 102, 103, 121, 132, 138, 153, 154, 202, 233
Airbus A380, 45
airplane, 56, 190
airport, 181, 190, 191
Al Sultan Abdullah Royal Tiger Reserve, 162
Alexander-Arnold Trent, 39
algae, 137, 192, 193
Algeria, 148, 230
al-Ghazali, 220

Ali, 160
Alps, 63
aluminium, 139, 146, 147
   Press Metal Aluminium, 146
American, 14, 45, 47, 113, 192, 207, 208, 214, 233, 236, 247
Amirudin Shari, 71
Amnesty International, 197
Ampang river, 82, 83
Andes, 238
Andina, Acción, 238
Antarctic Treaty, 204
Antarctica, 203, 204
Anthropocene, 9, 25
Anwar Ibrahim, 2, 17, 37, 42, 49, 86, 108, 121, 158, 186, 205, 220, 239
aquifers, 66, 116
Arab News, 231, 232
Arctic, 142, 148
Ardern, Jacinda, 1
Argentina, 204
   Buenos Aires, 69
Armizan Mohd Ali, 110

arsenic, 63, 66
ASEAN, 52, 54, 55, 56, 59, 152, 153, 154, 155, 156, 217, 245, 246
  AATHP, 154, 156
  Sub-Regional Ministerial Steering Committee on Transboundary Haze Pollution, 152
ASEAN Power Grid, 54, 55, 56, 59
Asian brown cloud, 98
Asian Development Bank, 215
Asian Financial Crisis, 35
Asian Infrastructure and Investment Bank, 215
asthma, 4, 100, 155
Atlantic, 56, 139
Australia, 25, 50, 51, 52, 54, 58, 150, 166, 191, 204
Ayni and Minka, 238
Azhar Azizan Harun, 109
Azman Mokhtar, 11

B'Tselem, 232
Bakun dam, 147
balik kampung, 61, 62, 94, 95
Balui river, 35
Banerjee, Abhijit V., 38, 206
Bangladesh, 139, 170, 209, 210
  Cox's Bazar, 209

Baniwa, 180
bantengs, 131, 164
Barber, Todd, 192
Barefoot College, 246
Bastille, 238
Batang Ai, 50
batik, 109, 110, 111
bats, 169, 170, 176
battery storage, 40, 54
Batu dam, 80
Bedouin, 94
Belgium, 58
Belize, 246
Belum–Temenggor forest complex, 163
benzene, 220
bicycles, 96, 100, 113, 117, 124
Biden, Joe, 211, 216
Big Cat Alliance, 175
biodiesel, 182
biodiversity, 4, 6, 14, 26, 61, 86, 97, 130, 157, 163, 176, 178, 187, 194, 202, 207, 233, 246, 247
biomass, 39, 42, 43, 45, 46, 47, 93
Birch, Ernest Woodford, 163
Black, James F, 234
Blake, William, 101
blanket subsidies, 33, 34, 37, 53, 122, 127, 205, 206, 211

BMW, 212
BN, 69
Boomitra, 238
bottled water, 63, 67
Boutros-Ghali, Boutros, 61
BP, 137
Brazil, 46, 157, 167, 186, 210, 224
  Rio de Janeiro, 112, 165, 229
Bretton Woods, 215, 229
Brexit, 9
BRT, 125
Brunei, 152, 164
Buddhists, 13
Burgess, Anthony, 169
Bursa Malaysia, 97
  BCX, 129, 130
buses, 100, 123, 124, 125
bus, 117, 124, 125
  GoKL, 125
  mini bus, 124, 125

C40 Cities network, 117
Caen, Herb, 93
calcium, 66
California, US, 29, 142, 148, 211
Calvinists, 13
Cambodia, 54, 153
Cambridge Central Mosque, 12
Canada, 42, 135, 148, 155, 178, 194, 210, 224
  Montreal, 178, 229
Canada–US Air Quality Agreement, 155
cancer, 4, 100
CAP, 25, 135
capitalism, 9, 25, 70, 94, 202, 213, 234
Capone, Al, 160, 161
Captain Planet and Planeteers, 229
carbon credit, 130, 131
carbon market, 129, 131, 132
carbon pricing, 132
carbon sink, 129, 167, 192, 246
Carney, Mark, 120, 194
cars, 32, 51, 96, 97, 100, 101, 102, 103, 121, 122, 123, 124, 125, 126, 128, 143, 157, 212, 230, 243
Carson, Rachel, 19, 25, 159
Catholic, 12
CATS, 173
CBAM, 217
CCC, 51
CCS, 27
cement industry, 58, 147
CFCs, 229, 230
Chain Reaction Research, 183
Chang, Ha-Joon, 203
charcoal, 168, 184
Chee Yoke Ling, 229
Chengal, 163
Chile, 35, 204

China, 43, 50, 58, 100, 101, 114, 116, 121, 122, 125, 126, 137, 141, 146, 147, 169, 170, 186, 190, 191, 193, 210, 215, 223, 237, 244
   Beijing, 100, 101, 115, 121
   Guilin, 223
   Han Dynasty, 13
   Kunming, 178
   Shanghai, 126
   Zhengzhou, 116
Chini lake, 183, 190
Choc, Cristina, 246
Choc, Miriam, 246
Choco, Florentina, 246
chocolates, 181, 182
cholek, 241
cholera, 62
Christianity, 12, 193
climate adaptation, 62, 120, 156, 157, 167
   National Adaptation Plan, 157
   PNBCAP, 157
Climate Change Consultative Panel, 221, 222
Climate Club for Foreign Ambassadors, 185
climate finance, 218
climate justice, 15, 201, 205, 213, 216, 219
climate migration, 209, 210
climate mitigation, 27, 156
climate reparations, 212, 214
Club of Rome, 233
coal, 29, 32, 36, 40, 42, 43, 47, 52, 53, 58, 94, 100, 101, 147, 151, 207, 218, 219
   mothballing, 42
coastal erosion, 27, 64, 167, 185, 192, 241
coconut, 21, 104
Cold War, 86, 204, 236
Coldplay, 76, 77
Colombia, 122, 231
   Bogota, 122
common but differentiated responsibilities, 203
community rangers, 178, 179
Confucian, 13
congestion charge, 102
Congo, 162, 231
conjunctivitis, 155
consumerism, 208
Convention on Biological Diversity, 178
Cool Biz campaign, 110
COP 15, 178, 219
COP 26, 207, 214, 218
COP 27, 214
COP 28, 14, 42, 43, 181, 207, 215, 216, 218, 223, 231, 232, 240
Coral Guardian, 193
corals, 165, 167, 168, 190, 191, 192, 193
CORSIA, 131

Costa Rica, 132, 187
Covid-19, 6, 120, 152, 169, 178, 194, 206, 207, 212, 214, 228, 237, 247, 248
crocodiles, 78
crown shyness, 185
Cullis-Suzuki, Severn, 224
cycling, 97, 149

Damansara river, 82
Danum Valley, 163
danwei, 100
DAP, 70, 144
Darwin, Charles, 160
Dasgupta Review, 194
Deepavali, 98
deer, 162, 173
deforestation, 5, 106, 129, 131, 156, 164, 167, 168, 169, 170, 175, 182, 183, 187, 214, 238
Delta Programme, 242
Denmark, 86, 219
diarrhoea, 149
DID, 142
diesel, 33, 48, 52, 96, 122, 124, 126, 128, 133, 182, 197
dipterocarp trees, 164
Djibouti, 187
DOE, 24, 190, 199
dolphins, 7, 185
Doughnut Economics, 201, 202
droughts, 12, 63, 90, 209, 210, 217

Duflo, Esther, 38, 39, 206
durian, 64, 169
Dusun, 21, 22
dzud, 209

early warning systems, 86, 156, 158
Earth Summit, 165, 203, 224, 229
Earthshot Prize, 1, 237, 238
East river, 86
EC, 42
ecocide, 232
ecofascism, 207
ecotourism, 77, 185, 187
EDF, 45
Education Infrastructure for Resilience Project, 243
EECA, 41, 109
EFT, 186
Egypt, 11, 47
EIA, 113, 189, 191
Ekosetiawangsa, 117
El Niño, 150, 151, 153, 170, 192
electric buses, 100, 123, 124
electric lorries, 51, 124
electric motorcycle, 127
electric vehicles, 51, 100, 101, 102, 121, 122, 124, 126, 127, 128, 206, 211, 237, 245
electricity tariff, 33, 34, 36, 71, 127

feed-in-tariff, 46
GET, 53
ICPT, 35, 36, 37, 53
elephants, 7, 131, 163, 164, 171, 172, 173
Elisara, Chris, Dr, 15
Emissions Trading Scheme, 132, 133
energy trilemma, 31, 52
Environmental Quality Act, 23, 24
environmentalism, 10, 25
EPA, 24, 25
EPSM, 25
EPU, 67
ESG, 120, 219, 220
Ethiopia, 210
EU, 4, 102, 132, 197, 210, 214, 215, 217, 218, 219, 243
European Investment Bank, 215
Exxon, 234, 241

FAM, 174
fatalism, 242
FatHopes Energy, 137
FDRS, 153
Federal Constitution, 67, 69
FELDA, 181
Financial Times, 220
fireflies, 114
First Nation Iroquois Confederacy, 247
First World War, 56, 213

floods, 5, 6, 27, 62, 63, 64, 76, 79, 80, 81, 82, 83, 84, 85, 86, 88, 89, 90, 112, 115, 116, 139, 141, 156, 157, 158, 206, 209, 214, 231, 240, 241, 242
flash floods, 76, 112, 115, 139
hazard maps, 156, 158
mitigation, 64, 79, 81, 83, 115, 242
tidal, 82, 86, 240
forests, 5, 6, 13, 20, 21, 22, 23, 26, 27, 77, 90, 104, 106, 111, 112, 113, 114, 129, 130, 131, 135, 151, 153, 155, 156, 162, 163, 164, 165, 166, 167, 170, 171, 172, 173, 174, 176, 177, 178, 179, 180, 181, 183, 184, 185, 186, 187, 188, 189, 190, 194, 195, 209, 214, 242, 244, 245
Amazon, 112, 132, 162, 176, 180, 194
plantations, 189
reserves, 23, 77, 113, 130, 131, 135, 162, 167, 184, 189, 190
Forest Management Certification, 186
fossil fuel, 8, 12, 29, 32, 33, 37, 39, 42, 44, 46, 52, 57,

94, 99, 119, 121, 122, 128, 133, 165, 166, 187, 198, 199, 200, 207, 210, 218, 219, 233, 234, 245
FPIC, 130
France, 42, 43, 56, 96, 186, 197, 204, 212, 216
   Paris, 4, 50, 69, 96, 178, 211, 244
Freakonomics, 230
Friedman, Thomas, 48
FRIM, 114, 184, 185
frogs, 177
fuel-cell
   vehicles, 101
   cars, 124, 128

Galain, Ramón Méndez, 45, 46
Gandhi, Mahatma, 197, 202
gaurs, 163, 173
Gayanashagowa, 247
GBI, 106, 107
Geely, 121
Germany, 43, 56, 208, 213, 216
   Berlin, 69
gibbons, 164
Gilets Jaunes, 197, 212, 216
glaciers, 61, 188, 238
global warming, 19, 131, 147, 149
Gold Standard, 129

Gombak river, 82
Gore, Al, 223
Great Green Wall, 187
Great Pacific Garbage Patch, 142
Great Smog, 94, 101, 102
Great Stink, 94
Greece, 148
green buildings, 106, 107
Green Climate Fund, 157
Green Revolution, 202
Greenpeace, 20, 182
greenwashing, 105, 131, 242
greywater, 73, 106
grid, 30, 31, 40, 41, 47, 48, 51, 53, 54, 55, 56, 59, 74, 129, 150
groundwater, 63, 65, 66, 84, 85, 134, 168
gum arabic, 188
Gummer, John, 51
gun lobby, 241
Guterres, Antonio, 148

Habeck, Robert, 208
Han river, 79
haze, 5, 7, 105, 120, 151, 152, 153, 154, 155, 156, 170, 235
health, 5, 9, 26, 38, 52, 57, 92, 98, 103, 107, 111, 136, 149, 150, 157, 230
heart problems, 100

heat islands, 94, 104, 105, 157
heat pumps, 12, 198
Heath, Edward, 24
heatstroke, 149
heatwave, 110, 148, 149, 150, 151
Helm, Dieter, 166, 198
Hidalgo, Anne, 96
highway, 79, 83, 123, 124, 125, 172, 173, 200
Hijau, 222, 227, 229
Hill, Fiona, 232
Hindenburg, 56, 57
Hindu Kush, 188
Hindus, 13, 98, 188
Hippocrates, 93
Hitler, Adolf, 213
hockey stick graph, 228
Honda, 128
hornbills, 163
Hult Prize, 136
Hurricane Ian, 87
Hurricane Irma, 176
Hurricane Katrina, 84
Hurricane Maria, 112, 176
hydroelectricity, 32, 35, 42, 45, 46, 50, 55, 127, 147, 165
hydrogen, 39, 56, 57, 58, 59, 124, 127, 128, 129, 147

ICAO, 131
Ice Age, 187
IMF, 215, 218
Inca, 238
Inconvenient Truth, 223
India, 20, 99, 101, 148, 151, 168, 170, 175, 186, 191, 197, 244, 246
   Delhi, 98
Indian Ocean, 5, 98
Indonesia, 5, 22, 54, 55, 56, 84, 152, 153, 154, 155, 156, 165, 167, 182, 183, 188, 193, 219, 245
   Aceh, 21, 168, 189
   Celebes, 159
   Jakarta, 84, 115, 116, 154
   Java, 115, 159, 189
   Kalimantan, 56, 153, 154, 168, 170
   Sumatera, 5, 56, 104, 154
   Sumatra, 153, 168, 170
Industrial Revolution, 14, 85, 93, 94, 101, 119, 166, 169, 177, 181, 210, 214, 228, 231
insects, 169, 176, 177, 185
Interceptor, 76, 77
International Ranger Award, 180
IPCC, 88
IPPs, 30, 31, 32, 35, 43
IRA, 216, 217
Iran, 32
Iraq, 32
Ireland, 67, 138, 157

iron, 63, 66, 147, 168
Islam, 10, 12, 166, 193, 197, 208
Islamic finance, 220
Ismail Abdul Rahman, 23
Israel, 232
Italy, 56, 138, 148
   Venice, 21
IUCN, 180, 182
IWK, 67, 68, 72, 73

Japan, 33, 40, 45, 47, 57, 58, 123, 128, 142, 210, 216, 219, 230
   Fukushima Nuclear Power Plant, 110
   Imperial Japanese Army, 240
JETP, 219
Johnson, Boris, 102, 243
Johor, Malaysia, 54, 151, 159, 171, 196, 220
Jokowi, 85, 219
Jordan, 10, 231

Kampung Baru, 103, 104, 105
KEADILAN, 2, 3, 80, 109, 144
Kedah, Malaysia, 30, 150, 168, 196
Keeling Curve, 233
Keeling, Charles David, 233
Kelantan river, 61, 64
Kelantan, Malaysia, 63, 64, 65, 66, 68, 69, 78, 79, 80, 81, 104, 126, 133, 149, 150, 160, 161, 163, 240
   Gua Musang, 63, 66, 174, 175
   Jeli, 160
   Kota Bharu, 61, 64, 80, 95, 125, 126, 149, 171
   Pasir Puteh, 64, 65
   Sabak beach, 240
Kennedy, John F., 25, 236, 237
keria, 201
Kerry, John, 42, 236, 237
Keynesian economics, 213
Khan, Abdul Ghaffar, 197
Khan, Imran, 188
Khan, Sadiq, 3, 4, 102, 243
Kim Kim river, 220
KL Car Free Morning programme, 97
Klang Gates dam, 80
Klang river, 76, 77, 78, 80, 82, 83
Kloth, 111
Kofe, Simon, 214
Koike, Yuriko, 110
Kongjian Yu, 114
Koon, Paul, 146
Korean War, 79
Kuala Lumpur Climate Action Plan 2050, 117
Kuala Lumpur, Malaysia, 7, 8, 33, 62, 63, 72, 76, 80, 82, 83, 91, 92, 97, 103, 104,

105, 106, 112, 113, 114,
116, 117, 124, 125, 126,
134, 141, 144, 145, 153,
154, 185, 196
Bukit Bintang, 105
Bukit Dinding, 8, 112
Bukit Kiara, 112, 113, 114
Kampung Baru, 103, 104, 105
Pudu, 105
Setiawangsa, 7, 8, 112, 113, 117, 178
Wangsa Maju, 112, 113, 117
Kulamba wildlife reserve, 183
Kunming–Montreal Global Biodiversity Framework, 178
KUTS, 127
Kyoto Protocol, 32, 157, 211, 234

La Niña, 7, 150, 152
Labuan, Malaysia, 71
landfills, 119, 129, 134, 135, 138, 139, 145
landslides, 5, 8, 156, 241
Laos, 55
LCCF, 116, 117, 129
lead, 229, 230
Lebanon, 12
Lee Myung-bak, 79
Leo Moggie Irok, 31
leopards, 163, 175, 183
leptospirosis, 62

limescale, 66
limestone, 66
Liverpool FC, 39, 71, 144, 145
Livingstone, Ken, 102
long-termism, 200, 221
Loss and Damage Fund, 214, 215, 216, 240
LRT, 117, 124
Ludosky, Priscilla, 197
lynxes, 175, 176

MacAskill, William, 200
Macron, Emmanuel, 42, 43
Madani Agenda, 239
Madrasah Ihya Ulumuddin, 64, 65
magnesium, 66
Mahathir Mohamad, Dr, 25, 33, 34, 70, 122, 123, 166, 190, 203, 204
Malacca, Malaysia, 20, 21, 73, 117, 128, 159, 163
malaria, 222
Malaysian Palm Oil Board, 151
Maliau Basin, 164
manganese, 63, 66
mangroves, 5, 85, 90, 115, 167, 168, 183, 184, 185, 192, 214
Mann, Michael E., 222, 228, 241, 242
maqasid sharia, 220
Mariana Trench, 143
marine park, 190, 191

Marshall Plan, 213, 214
Matang Mangrove Forest
 Reserve, 184
Maybank, 174
MCKK, 29, 108, 123, 148
meat, 136, 162, 192, 241
Mek Jah Ismail, 160, 161
Mencius, 13, 159
Menraq, 179, 180
Meranti, 163
Merbau, 163
mercury, 135
Metallica, 92
methane, 58, 129, 134, 220
Methodists, 220
METMalaysia, 153
Mexico, 210
micromobility, 97, 124
mineral water, 63
mining, 50, 76, 135, 167, 185,
 204, 238
Mississippi river, 84
molluscs, 191
Mongolia, 209
monorail, 124
monsoon, 61, 78, 141
Moonlight Project, 40, 57
Moore, Charles, 142
Moral Money Summit, 220
MRT, 122
Muda river, 196
Musa Hitam, 23
Musk, Elon, 121
Myanmar, 209

Nancy Shukri, 109, 110
National Depletion Policy, 32
National Forestry Act, 167,
 189
National Forestry Council, 167
National Forestry Policy, 167
National Haze and Dry
 Weather Main Committee,
 150
National Policy on
 Biodiversity, 26
National University of
 Malaysia, 59, 184
National Wildlife Recovery
 Centre, 161
nationalization, 35, 68
Native American, 247
natural gas, 32, 35, 40, 41, 52,
 58, 201
nature-based solutions, 26, 89,
 116, 158, 184
NDC, 50, 239
Needleman, Herbert, 230
Nenggiri river, 64
NEP, 35
Netherlands, 62, 242
 Amsterdam, 84
NETR, 42, 49
net-zero, 9, 26, 27, 29, 39,
 48, 49, 51, 54, 55, 59,
 88, 146, 198, 199, 200,
 239, 245
New Zealand, 1, 25, 145, 204,
 208

Nik Ahmad Nik Hassan, 19
Nik Nazmi Nik Ahmad, 19, 33, 216, 235
nimbyism, 52, 241
Nipah virus, 169, 170
nitrogen, 96, 99, 100, 103
North West Water, 67
North–South Expressway, 123, 125
Norway, 138
NRECC, 2, 3, 16, 19, 23, 31, 36, 41, 44, 45, 48, 69, 72, 75, 86, 106, 165, 204, 235, 239
NRES, 2, 16, 19, 26, 72, 106, 117, 133, 157, 204, 222
NRW, 66, 71, 74, 75
nuclear, 40, 45, 110
Nurul Izzah Anwar, 109

Obama, Barack, 211
Ocean Cleanup, 76, 77, 78
Official Secrets Act, 154
off-river storage, 83
oil palm, 46, 64, 78, 82, 183
   palm oil, 46, 151, 153, 181, 182, 183
OneRepublic, 238
OPEC, 31
open burning, 98, 99
Orang Asli, 20, 63, 64, 66, 104, 173, 174, 178, 179, 180, 183
   Cheq Wong, 20
   Jahai, 179
   Jakun, 183
   Mah Meri, 20
   Seletar, 20
   Semelai, 20
   Temiar, 20
orangutans, 131, 164, 183
otters, 78
Ou Hongyi, 223
Oxfam, 211
ozone layer, 193, 202, 227, 229, 230

Pacific, 5, 47, 142, 143, 150, 165
Pacific Ocean, 5, 142
Pacino, Al, 160
PACOS, 131
PADU, 37
Pahang, Malaysia, 7, 71, 84, 135, 151, 161, 162, 172, 196, 240
   Lanchang, 162
   Tioman Island, 190, 191
Paisley, Bob, 71
Pakistan, 12, 88, 98, 188, 197, 214
   Khyber Pakhtunkhwa, 188
Palestine, 231, 232
   Gaza, 231, 232
   West Bank, 232
Pantamera, 138
Papua New Guinea, 159, 183
Paris Agreement, 4, 50, 178,

211, 244
Parliamentary Special Select Committee on the Environment, Science and Plantations, 240
particulate matter, 96, 99, 101
PAS, 144, 227
Pearce, Fred, 89
peat, 5, 7, 151, 152
peatlands, 151, 153
PEFC, 186
Penang Climate Board, 157
Penang, Malaysia, 22, 23, 73, 144, 145, 157, 168, 196
Penchala river, 78
Pengkalan Datu river, 78
Perak river, 30
Perak, Malaysia, 7, 161, 162, 163, 168, 173, 180, 184, 196
  Ipoh, 30, 125
  Kampung Belum, 163
  Kuala Kangsar, 30, 124, 125
  Sungkai, 161, 162
Perlis, Malaysia, 71, 196
Perodua, 123
Peruvian Amazon, 132
pesticides, 25, 78, 177
Petro, Gustavo, 122
petrol, 32, 48, 96, 102, 121, 122, 127, 182, 196, 197, 205, 206, 229, 230
petroleum, 40, 42, 47, 48, 57, 123, 128, 133, 144, 201, 233, 245
Petronas, 31, 59, 103, 105, 134, 137
Petronas Twin Towers, 103, 105, 134
Petros, 128
Pfeiffer, Michelle, 160
PH, 2, 3, 125, 203
Pham Minh Chinh, 42
Piketty, Thomas, 216
Pires, Tomé, 21
Plant for Pakistan, 188
plastics, 119, 137, 139, 140, 142, 143, 144, 145, 146, 201
  microplastics, 143, 145
  nanoplastics, 143
pollution, 4, 7, 57, 61, 69, 76, 78, 83, 90, 93, 94, 95, 96, 98, 99, 100, 101, 102, 103, 105, 112, 114, 120, 121, 132, 138, 141, 153, 154, 168, 198, 202, 207, 222, 242
  air, 93, 96, 98, 101, 102, 103, 121, 132, 138, 153, 154, 202
Pope Francis, 12, 13
  Laudate Deum, 12, 13
  Laudato Si', 12
Porsche, 212
Portugal, 186
PPBM, 227
PR, 3, 25, 144

PRABN, 86
Presbyterians, 13
Prince William, 237
privatization, 25, 33, 34, 35, 64, 67, 68, 69, 70, 90
Proton, 121, 123
Przewalski's horse, 176
PSI, 153
public transport, 102, 117, 122, 123, 124, 126, 206
Puerto Rican Amazon, 176
Puerto Rico, 177
pumped energy storage, 51
Putrajaya, Malaysia, 106, 116, 134, 145, 196
Pyle, Barbara, 229

Qatar, 52, 73
Q-Cells, 47, 50
Quakers, 220
Quran, 10

Rafizi Ramli, 36, 42, 49, 55
rail, 123, 124, 125, 126
   ECRL, 125, 126
   ETS, 125, 126, 133
   high-speed rail, 126
   Jungle Railway, 126
   Malayan Railway, 123
   Shinkansen, 126
rainwater, 12, 73, 80, 107, 115, 116, 206

rainwater harvesting, 73, 107, 206
Ramadan, 133, 134
Raworth, Kate, 201, 202
Razzouk, Assaad, 139
RE100, 54
Reagan, Ronald, 34, 70
recycling, 73, 111, 120, 134, 137, 138, 139, 145, 147, 220
reef ball, 192
reforestation, 112, 132, 167, 185, 187
Reman Kingdom, 163
Rhino and Forest Fund, 183
rhinos, 170, 171, 183
Rice University, 236
Rimau, 180
Rohingya, 209
Roman, 93, 230
Royal Belum state park, 162
rubber, 63, 82, 112, 114, 160, 161, 167, 181
Russia, 46, 52, 110, 176, 207, 210, 232
   Siberia, 148
Russia–Ukraine war, 46, 110, 176, 207, 232
Rwanda, 139, 140

Sabah, Malaysia, 5, 21, 22, 26, 51, 56, 59, 79, 110, 130, 150, 163, 164, 165, 170, 171, 180, 183, 240

Kuamut Rainforest Conservation Project, 130
Sahabat Alam Malaysia, 229
Sahara desert, 187
Salahuddin Ayub, 107, 108
Sanchez, Pedro, 110
sanitation, 87, 232
Santana, 92
Sarawak Law, 160
Sarawak, Malaysia, 26, 30, 35, 50, 51, 54, 56, 59, 86, 127, 128, 137, 143, 147, 151, 154, 159, 160, 164, 180, 193, 240
    Bintulu, 59
satellite, 47, 98, 154, 236
Saudi Arabia, 218, 231
    Mecca, 93, 94
Scarface, 160, 161
Scholz, Olaf, 197, 198
Schwarzenegger, Arnold, 29
SDGs, 201
Second World War, 6, 24, 86, 163, 200, 213, 240
Selangor Maritime Gateway, 76
Selangor river, 76, 83
Selangor, Malaysia, 7, 8, 47, 48, 50, 70, 71, 76, 80, 83, 112, 125, 134, 135, 144, 147, 151, 152, 181, 196
    Gombak, 82, 104
    Kelana Jaya, Petaling Jaya, 7, 135
    Klang, 20, 73, 76, 80, 82, 124, 125, 126, 135, 139, 141
    Petaling Jaya, 78, 135
    Puchong, 146
    Shah Alam, 135
    Subang Jaya, 135
    Sungai Besar, 151
    Sunway, 125
Senegal, 42, 187, 188
Senoi Praaq, 178
SESB, 150
sewage, 2, 19, 70, 72, 73, 76, 78, 79, 120
sewerage, 64, 67, 68, 72, 73
sharia, 129, 220
Shuman, Frank, 47
Singapore, 1, 23, 25, 33, 35, 54, 55, 73, 137, 138, 139, 152, 153, 154, 155, 156, 159, 163, 196, 238
    Semakau island, 138
    Transboundary Haze Pollution Act, 155
Skjern river, 86
SMART Tunnel, 83
SMEs, 37, 219
Smith, Adam, 202
solar, 37, 39, 40, 42, 44, 45, 46, 47, 48, 50, 51, 52, 53, 55, 57, 59, 72, 75, 76, 104, 106, 107, 117, 146, 199, 206, 211, 234, 237, 245, 246

Solidaridad, 182
Solomon Islands, 165
South Africa, 231
  Johannesburg, 69
South China Sea, 190, 193
South Korea, 33, 79
  Seoul, 79
Soviet Union, 86, 198, 204, 229, 236
Spain, 176, 197
SPAN, 65, 66, 74, 150
Speth, Gus, 14, 231
sponge city, 114, 115, 116
Sri Lanka, 153, 168
ST, 41, 49
Starmer, Keir, 243
steel industry, 58, 146, 147
Stevens, Cat, 12
Storm Desmond, 85
Storm Eva, 85
Storm Frank, 85
Straits of Malacca, 21, 128
stroke, 4, 100, 149, 230
Sultan Abdullah, 161, 162, 240
Sultan Mizan Antarctic Research Foundation, 204
sun bears, 164, 173, 183
Sungai Rasau, 83
Sunshine Project, 57
sustainable aviation fuel, 136, 137
SUV, 96

Sweden, 24, 137, 138
Switzerland, 25
Syria, 242

Tabin wildlife reserve, 183
Taiwan, 33
Taman Negara, 165, 172
tapirs, 7, 163
targeted subsidies, 35, 38, 127, 133, 206
Tengku Hassanal Ibrahim Alam Shah, 161, 162, 240
Terengganu, Malaysia, 30, 151, 170, 173, 241
  Rantau Abang, 170, 171
  Tok Jembal beach, 241
Tesla, 121, 128
Thailand, 30, 46, 55, 149, 152, 162, 163, 168
Thames river, 70, 94
Thames Water, 68, 69, 70
Tharoor, Shashi, 195
Thatcher, Margaret, 34, 68, 70
The Edge, 23, 38
The Limits to Growth, 233
the Philippines, 5, 55, 153, 165
The Police, 71
The Theory of Moral Sentiments, 202, 203
Think City, 105
Thunberg, Greta, 223

tigers, 160, 161, 162, 163, 171, 173, 174, 175, 178, 179, 180
timber, 23, 26, 35, 167, 189
TM, 113
TNB, 30, 31, 33, 35, 36, 37, 44, 45, 53, 59, 146
tobacco lobby, 241
tody, 177
Toyota Mirai, 128
traffic jams, 96, 105, 125
Trail, Armitage, 160
Treaty of Versailles, 213
Trump, Donald, 9, 195, 207, 211
tsunami, 156, 168, 188
Tuan Ibrahim Tuan Man, 66
Tuan Tabal, 64
TUG, 136
Tunisia, 148
Türkiye, 12, 148, 231, 242
Turtle Hospital, 191
turtles, 7, 170, 171, 191
Tuvalu, 214

UAE, 73, 216
   Dubai, 14, 181, 182, 240
UK, 24, 25, 34, 40, 42, 43, 44, 45, 51, 53, 56, 66, 67, 68, 69, 101, 102, 103, 110, 122, 129, 138, 194, 204, 216, 243
   England, 44, 45, 67, 69, 85, 194
   Glasgow, 218
   London, 3, 4, 39, 66, 89, 94, 101, 102, 103, 120, 139, 159, 162, 166, 194, 196, 198, 201, 202, 203, 206, 211, 237, 242, 243
   Northern Ireland, 67
   Scotland, 67, 128
   Wales, 67, 238
Ukraine, 46, 110, 176, 207, 232
ULEZ, 102, 103, 243
UN, 5, 14, 24, 25, 61, 70, 86, 87, 89, 130, 140, 148, 149, 165, 178, 204, 230, 239
UN Biodiversity Conference, 178
UNCLOS, 203
UNDP, 186
UNFCCC, 32, 166
University of Notre Dame, 231
upper-respiratory tract infection, 155
urbanization, 94, 95, 96
Uruguay, 45, 46
US, 12, 24, 31, 34, 42, 50, 52, 57, 66, 86, 87, 101, 115, 121, 132, 146, 155, 156, 157, 160, 166, 195, 204, 207, 208, 210, 211, 213, 215, 216, 217, 218, 219, 232, 234, 236, 237, 239, 244
   California, 29, 142, 148, 211
   Chicago, 237

Miami, 157
New Orleans, 84
New York, 10, 56, 85, 86, 87, 153, 200, 237, 244
San Francisco, 92, 93
Usman Awang, 92

van Dijk, Virgil, 39
Vantage RE, 44, 45
vegetable oil, 181, 182, 183
Verra, 129, 130
Vietnam, 42, 46, 53, 55, 153
Vision 2020, 121, 123
Volkswagen, 212
Volvo, 121
von Zeppelin, Ferdinand, 56

Walker, Peter, 24
Wallace, Alfred Russel, 159, 160
Wan Gang, 100, 101
Wan Junaidi Tuanku Jaafar, 155
waqf, 73, 74, 188, 189
waste, 6, 9, 35, 46, 73, 76, 77, 78, 88, 90, 94, 95, 117, 119, 127, 133, 134, 135, 136, 137, 138, 139, 141, 144, 145, 156, 176, 224, 237
water pollution, 69, 94, 120
water recycling, 73
Water Sector Transformation 2040, 89
water tariff, 70, 71, 72, 73, 205
  tariff-setting mechanism, 72
Wessex Water, 68
wetlands, 86, 115, 116, 197
WHO, 4, 169
Wilde, Oscar, 16
wildlife department, 161, 174, 178
wildlife reserve, 183, 190
wind turbines, 42, 45, 46, 237
WMO, 4, 148
Wong, Hailey, 136
World Bank, 75, 133, 209, 215, 219, 243
World Evangelical Alliance, 15
World Resources Institute, 183, 213
WTE, 137, 138, 139
WWF, 4, 25, 182, 183

Xi Jinping, 115

Yayasan Hijau Malaysia, 222
Yemen, 231
Yeo Bee Yin, 155
Yinson GreenTech, 128
YTL, 68

Zainal Abidin, 227, 229
Zakaria, Doliwura, Dr, 15
Zirkelbach, Bette, 191

Select Bibliography

'A Study of Privatization in Malaysia', *JICA*, 1999, https://openjicareport.jica.go.jp/pdf/11634078.pdf [Accessed August 9, 2023].

Banerjee, Abhijit V. and Esther Duflo, *Good Economics for Hard Times*. London: Penguin, 2020.

Butfield, Colin and Jonnie Hughes, *Earthshot: How to Save the Planet*. London: John Murray, 2021.

Carney, Mark, *Value(s): Climate, Credit, Covid and How We Focus on What Matters*. Revised and updated edition. London: William Collins, 2021.

Carson, Rachel, *Silent Spring*. Boston: Houghton Mifflin Company, 1962.

Cooke, Fadzilah Majid and Adnan A. Hezri, 'Environmental Activism in Malaysia: Struggling for Justice from Indigenous Lands to Parliamentary Seats', in *Environmental Movements and Politics of the Asian Anthropocene*, Jobin, Paul, Ho Ming-sho, and Michael Hsiao Hsin-Huang (eds.), Singapore: ISEAS, 2021, pp. 203–231.

Crouch, David, *Bumblebee Nation: The Hidden Story of the New Swedish Model*. Stockholm: Karl-Adam Bonniers Stiftelse, 2018.

Dasgupta, Partha, 'The Economics of Biodiversity: The Dasgupta Review', *HM Treasury*, 2021, https://assets.publishing.service.gov.uk/media/602e92b2e90e07660f807b47/The_Economics_of_Biodiversity_The_Dasgupta_Review_Full_Report.pdf [Accessed February 20, 2024].

Friedman, Thomas, *Hot, Flat, and Crowded: Why We Need a Green Revolution – and How It Can Renew America*. New York: Farrar, Straus and Giroux, 2008.

Goh Chun Seng and Potter, Lesley, *Transforming Borneo: From Land Exploitation to Sustainable Development*. Singapore: ISEAS, 2023.

Helm, Dieter, *Net-zero: How We Stop Causing Climate Change*. London: William Collins, 2021.

'Malaysia Energy Transition Outlook', *IRENA*, 2023, https://www.irena.org/Publications/2023/Mar/Malaysia-energy-transition-outlook [Accessed August 9, 2023].

Jobin, Paul, Ho Ming-sho and Michael Hsiao Hsin-Huang (eds.), *Environmental Movements and Politics of the Asian Anthropocene*. Singapore: ISEAS, 2021.

Khan, Sadiq, *Breathe: Tackling the Climate Emergency*. London: Hutchinson: Heinemann, 2023.

Khor, Martin and Raman, Meenakshi, *A Clash of Climate Change Paradigms: Negotiations and Outcomes at the UN Climate Convention*. Penang: Third World Network, 2020.

Lee Hoseung and Romero, Jose (eds.), 'Synthesis Report. A Report of the Intergovernmental Panel on Climate Change. Contribution of Working Groups I, II and III to the Sixth Assessment Report of the Intergovernmental Panel on Climate Change', *IPCC*, Geneva, 2023, https://www.ipcc.ch/report/ar6/syr/downloads/report/IPCC_AR6_SYR_FullVolume.pdf [Accessed September 10, 2024].

Linden, Eugene, *Fire and Flood: A People's History of Climate Change from 1979 to the Present*. London: Penguin, 2023.

MacAskill, William, *What We Owe the Future*. New York: Basic Books, 2022.

Mann, Michael E, *The New Climate War: The Fight to Take Back Our Planet*. London: Scribe Publications, 2022.

Mitchell, Timothy, *Carbon Democracy in the Age Oil*. 2023 edition. London: Verso, 2011.

Mosley, Stephen, 'Environmental History of Air Pollution and Protection', in *The Basic Environmental History*, Agnoletti, Mauro and Serneri, Simone Neri (eds.), Cham: Springer International Publishing, 2014, 143–169.

Nik Nazmi Nik Ahmad, *In the Public Service: The Life of My Father, Nik Ahmad*. Kuala Lumpur: Pusat Sepakat, 2018.

Pearce, Fred, *When the Rivers Run Dry: The Global Water Crisis and How to Solve it*. London: Granta Publications, 2019.

Pope Francis, 'Laudate Deum', *Vatican*, October 4, 2023, https://www.vatican.va/content/francesco/en/apost_exhortations/documents/20231004-laudate-deum.html [Accessed October 7, 2023].

Raworth, Kate, *Doughnut Economics: Seven Ways to Think Like a 21st Century Economist*. London: Penguin, 2017.

Razzouk, Assaad, *Saving the Planet without the Bullsh*t: What They Don't Tell You About the Climate Crisis*. London: Atlantic Books, 2022.

Smith, Adam, *The Theory of Moral Sentiments*. Stewart edition. London: Henry G. Bohn, 1853, 27, https://oll.libertyfund.org/titles/smith-the-theory-of-moral-sentiments-and-on-the-origins-of-languages-stewart-ed [Accessed February 14, 2024].

Thunberg, Greta (ed.), *The Climate Book*. London: Allen Lane, 2022.

van Wyhe, John and Drawhorn, Gerrell M., '"I am Ali Wallace": The Malay Assistant of Alfred Russel Wallace', *Journal of the Malaysian Branch of the Royal Asiatic Society*, vol. 88, 1 (308), 2015, pp. 3–31, https://www.jstor.org/stable/26527691 [Accessed January 21, 2024].

Wallace, Alfred Russel, 'On the Law Which Has Regulated the Introduction of New Species (1855)', *Alfred Russel Wallace Classic Writings*, Paper 2. 2009, https://digitalcommons.wku.edu/dlps_fac_arw/2 [Accessed February 14, 2024].

Wallace, Alfred Russel, *Tropical Nature and Other Essays*. London and New York: Macmillan & Co, 1878.
Yeo Bee Yin, *The Unfinished Business*. Kuala Lumpur: REFSA, 2022.
Yergin, Daniel, *The New Map: Energy, Climate and the Clash of Nations*. New York: Penguin Press, 2021.

Acknowledgements

The names involved in this project would be too many to mention, but here are a few of those who have played pivotal roles in my success as a policymaker and legislator, as well as in making this book possible.

I first communicated with Prime Minister Anwar Ibrahim by letter when I was an undergraduate in London, and he was imprisoned in Sungai Buloh, prior to his release in 2004. Subsequently, he gave me the opportunity to contest in Selangor and continued to support my rise in politics at a young age until he trusted me to be NRECC and NRES minister. The Prime Minister has also made my job easier with his immense support for energy transition and ensuring our flood mitigation plans become a reality. He has made sustainability a cornerstone of his Madani Agenda. His grit and determination to never give up, even after more than a decade of being imprisoned and facing many disappointments, are an inspiration not just to Malaysians but democrats worldwide.

Economic Minister Rafizi Ramli is the person who recruited me into politics as a nineteen-year-old. The public knows about his courage in speaking truth to power, raking in jail sentences for exposing corruption. But, as a friend, I can also testify to his comradeship and loyalty. We used to work closely when he was KEADILAN's strategic director and I was communications director. In government, we collaborated closely, as the economic ministry oversees overall energy policy while the then NRECC used to oversee electricity supply and renewable energy.

To the officers in the ministry office: Syazwan, my political secretary who is also an advisor on communications and engagement with civil society. Mega, my senior private secretary, who is not only the government servant that coordinates our work with the civil service but is probably the most passionate tree-hugger in Putrajaya. Akmal, Prasanth, Murnie, Faiz, Alyaa, Nadirah, Yasmin, Johan, Arif, Hazrol, and Irfan ensure that we survive the tough and challenging aspects of the job by keeping a fun office! I was happy to work with a talented and youthful team that supported my work and helped me to settle down in my new role. All the photos in the book are from my own collection or Mohd Hazrol Zainal's.

As the son of a civil servant, I totally appreciated the role played by many government officers that educated me about the role of the ministry and then enabled new policies and reforms to take place to align the ministry's policies with the new Unity Government. To Secretary General Dr Ching Too Kim, with his deep passion and knowledge about water, as well as everyone else at the ministry—I thank you. I was inspired to find civil servants at various levels of government who are just as passionate about environment and climate change, just like activists.

I am grateful to members of the civil society who directly shared their views for this book. Special mention goes to two marine biologists by training: Dr Yasmin Rasyid, who shared her research on the history of the environmental movement in Malaysia as well as her own experience with NGOs and the private sector, as well as my old friend Shahrinaz Maamor, who is a passionate scuba diver and has spent time working for various marine conservation efforts as well as corporate initiatives with regards to climate and coastal resilience. My MCKK senior Amin

Ramli is passionate about sustainability and was generous enough to spend time going through my book.

Keith Leong, Dr Idlan Rabihah Zakaria, Rahman Imuda, and Tunku Siti Ameerah have been my long-term friends who go through my work and offer criticisms and suggestions that make it better. Ewanina Effandie, one of the young Malaysians on the Climate Change Consultative Panel, also read through the draft and offered her views and ideas to ensure this book is accessible.

My parents have played a tremendous role in getting me where I am today. Other than providing me with the love and support to ensure my success, they taught me the importance of distinguishing between right and wrong and having a mind that is not only always curious but also critical of the information that it receives.

After I was appointed, Ilhan, my son, would snap photos of trees being felled and land being cleared for development and send me a WhatsApp message, 'You are not doing your job, Minister.'

It goes without saying that all mistakes are mine alone.